ASSESSING THE TMDL APPROACH TO WATER QUALITY MANAGEMENT

Committee to Assess the Scientific Basis of the Total Maximum
Daily Load Approach to Water Pollution Reduction

Water Science and Technology Board
Division on Earth and Life Studies

National Research Council

National Academy Press
Washington, D.C. 2001

NOTICE: The project that is the subject of this report was approved by the Governing Board of the National Research Council, whose members are drawn from the councils of the National Academy of Sciences, the National Academy of Engineering, and the Institute of Medicine. The members of the committee responsible for the report were chosen for their special competencies and with regard for appropriate balance.

Support for this project was provided by the U.S. Environmental Protection Agency under Cooperative Agreement No. X-82880401.

International Standard Book Number 0-309-07579-3

Assessing the TMDL Approach to Water Quality Management is available from the Water Science and Technology Board, 2101 Constitution Avenue, N.W., HA 462, Washington, D.C. 20418; (202) 334-3422; Internet <http://www.nap.edu>.

Copyright 2001 by the National Academy of Sciences. All rights reserved.

Printed in the United States of America.

THE NATIONAL ACADEMIES
Advisers to the Nation on Science, Engineering, and Medicine

National Academy of Sciences
National Academy of Engineering
Institute of Medicine
National Research Council

The **National Academy of Sciences** is a private, nonprofit, self-perpetuating society of distinguished scholars engaged in scientific and engineering research, dedicated to the furtherance of science and technology and to their use for the general welfare. Upon the authority of the charter granted to it by the Congress in 1863, the Academy has a mandate that requires it to advise the federal government on scientific and technical matters. Dr. Bruce M. Alberts is president of the National Academy of Sciences.

The **National Academy of Engineering** was established in 1964, under the charter of the National Academy of Sciences, as a parallel organization of outstanding engineers. It is autonomous in its administration and in the selection of its members, sharing with the National Academy of Sciences the responsibility for advising the federal government. The National Academy of Engineering also sponsors engineering programs aimed at meeting national needs, encourages education and research, and recognizes the superior achievement of engineers. Dr. William A. Wulf is president of the National Academy of Engineering.

The **Institute of Medicine** was established in 1970 by the National Academy of Sciences to secure the services of eminent members of appropriate professions in the examination of policy matters pertaining to the health of the public. The Institute acts under the responsibility given to the National Academy of Sciences by its congressional charter to be an adviser to the federal government and, upon its own initiative, to identify issues of medical care, research, and education. Dr. Kenneth I. Shine is president of the Institute of Medicine.

The **National Research Council** was organized by the National Academy of Sciences in 1916 to associate the broad community of science and technology with the Academy's purposes of furthering knowledge and advising the federal government. Functioning in accordance with general policies determined by the Academy, the Council has become the principal operating agency of both the National Academy of Sciences and the National Academy of Engineering in providing services to the government, the public, and the scientific and engineering communities. The Council is administered jointly by both Academies and the Institute of Medicine. Dr. Bruce M. Alberts and Dr. William A. Wulf are chair and vice chair, respectively, of the National Research Council.

COMMITTEE TO ASSESS THE SCIENTIFIC BASIS OF THE TOTAL MAXIMUM DAILY LOAD APPROACH TO WATER POLLUTION REDUCTION

KENNETH H. RECKHOW, *Chair*, Duke University, Durham, North Carolina
ANTHONY S. DONIGIAN, JR., AQUA TERRA Consultants, Mountain View, California
JAMES R. KARR, University of Washington, Seattle
JAN MANDRUP-POULSEN, Florida Department of Environmental Protection, Tallahassee
H. STEPHEN McDONALD, Carollo Engineers, Walnut Creek, California
VLADIMIR NOVOTNY, Marquette University, Milwaukee, Wisconsin
RICHARD A. SMITH, U.S. Geological Survey, Reston, Virginia
CHRIS O. YODER, Ohio Environmental Protection Agency, Groveport

NRC Staff

LEONARD SHABMAN, Visiting Scholar, Virginia Polytechnic Institute and State University
LAURA J. EHLERS, Study Director
M. JEANNE AQUILINO, Administrative Associate

WATER SCIENCE AND TECHNOLOGY BOARD

HENRY J. VAUX, JR., *Chair*, Division of Agriculture and Natural Resources, University of California, Oakland
RICHARD G. LUTHY, *Vice Chair*, Stanford University, California
RICHELLE M. ALLEN-KING, Washington State University, Pullman
GREGORY B. BAECHER, University of Maryland, College Park
JOHN BRISCOE, The World Bank, Washington, D.C.
EFI FOUFOULA-GEORGIOU, University of Minnesota, Minneapolis
STEVEN P. GLOSS, University of Wyoming, Laramie
WILLIAM A. JURY, University of California, Riverside
GARY S. LOGSDON, Black & Veatch, Cincinnati, Ohio
DIANE M. MCKNIGHT, University of Colorado, Boulder
JOHN W. MORRIS, J.W. Morris Ltd., Arlington, Virginia
PHILIP A. PALMER (Retired), E.I. du Pont de Nemours & Co., Wilmington, Delaware
REBECCA T. PARKIN, George Washington University, Washington, D.C.
RUTHERFORD H. PLATT, University of Massachusetts, Amherst
JOAN B. ROSE, University of South Florida, St. Petersburg
JERALD L. SCHNOOR, University of Iowa, Iowa City
R. RHODES TRUSSELL, Montgomery Watson, Pasadena, California

Staff

STEPHEN D. PARKER, Director
LAURA J. EHLERS, Senior Staff Officer
JEFFREY W. JACOBS, Senior Staff Officer
MARK C. GIBSON, Staff Officer
WILLIAM S. LOGAN, Staff Officer
M. JEANNE AQUILINO, Administrative Associate
PATRICIA A. JONES KERSHAW, Staff Associate
ANITA A. HALL, Administrative Assistant
ELLEN A. DE GUZMAN, Senior Project Assistant
ANIKE L. JOHNSON, Project Assistant
NORA BRANDON, Project Assistant
RHONDA BITTERLI, Editor

Preface

The Total Maximum Daily Load (TMDL) program, initiated in the 1972 Clean Water Act, recently emerged as a foundation for the nation's efforts to meet state water quality standards. A "TMDL" refers to the "total maximum daily load" of a pollutant that achieves compliance with a water quality standard; the "TMDL process" refers to the plan to develop and implement the TMDL. Failure to meet water quality standards is a major concern nationwide; it is estimated that about 21,000 river segments, lakes, and estuaries have been identified by states as being in violation of one or more standards. To address this problem, the U. S. Environmental Protection Agency (EPA) proposed an ambitious timetable for states to develop TMDL plans that will result in attainment of water quality standards. Given the reduction in pollutant loading from point sources such as sewage treatment plants over the last 30 years, the successful implementation of most TMDLs will require controlling nonpoint source pollution.

These two features, the ambitious timetable and nonpoint source controls, are probably the two most controversial of many issues that have been raised by those who have questioned the TMDL program. Behind and intertwined with these basic policy issues are important questions concerning the adequacy of the science in support of TMDLs.

In the last year, the TMDL program has become one of the most discussed and debated environmental programs in the nation, primarily because of the drafting of final rules for the program. These rules follow several years of intense activity, including the formation of a Federal Advisory Committee devoted to this topic. In October 2000, Congress suspended EPA's implementation of these rules until further information could be gathered on several aspects of the program. In particular, Congress requested that the National Research Council (NRC) examine the scientific basis of the TMDL program. In recognition of the urgent need to address water quality standard violations, Congress established an aggressive schedule for completion of the study that allowed only four months from start to finish—unprecedented for most NRC studies. The eight-member committee, constituted in January 2001, immediately conducted its first meeting. This three-day meeting included two days devoted to public comments and a third day focused on internal committee discussions. The ensuing three months was a period of intense activity filled with correspondence, writing, and two additional committee meetings.

The difficult challenges facing EPA and the states in the implementation of the TMDL program were immediately apparent to the committee. Because the committee faced a congressionally mandated deadline, a number of issues important to some stakeholders were not addressed comprehensively. These include bed sediment issues, atmospheric deposition, translating narrative standards into numeric criteria, and a full review of existing water quality models. Nonetheless, the committee found that substantial improvements can be made in a number of areas to strengthen the scientific basis of the TMDL program. Also of importance, the committee identified several policy issues that are restricting the use of the best science in the TMDL program. We urge Congress, EPA, and the states to give thoughtful attention to the recommendations made throughout this report so that resources can be more efficiently used to improve water quality.

We greatly appreciate the assistance of Don Brady and Françoise Brasier of the EPA Office of Water for their assistance in initiating the study and organizing the first committee meeting. We are also grateful to those who spoke with and educated our committee, including congressional staff, EPA scientists, state representatives, and the many individuals and organizations that submitted comments to the committee.

The committee recognizes the vital role of Water Science and Technology Board (WSTB) director Stephen Parker in making this study possible. The extremely short time period for this study created an enormous challenge for NRC study director Laura Ehlers, who was able to juggle her many responsibilities to keep us focused and provide invaluable assistance in crafting the text. Finally, it is fair to say that this study owes most thanks to Leonard Shabman (Virginia Polytechnic Institute and State University) who was working in the WSTB office as a visiting scholar during the study. Dr. Shabman's insight was invaluable; he added immensely to committee discussion and correspondence, and he

Preface

played a key role in drafting the text and developing the recommendations.

More formally, the report has been reviewed by individuals chosen for their diverse perspectives and technical expertise, in accordance with procedures approved by the NRC's Report Review Committee. The purpose of this independent review is to provide candid and critical comments that will assist the authors and the NRC in making the published report as sound as possible and to ensure that the report meets institutional standards for objectivity, evidence, and responsiveness to the study charge. The reviews and draft manuscripts remain confidential to protect the integrity of the deliberative process. We thank the following individuals for their participation in the review of this report: Richard A. Conway, consultant; Paul L. Freedman, Limno-Tech, Inc.; Donald R. F. Harleman, Massachusetts Institute of Technology (retired); Robert M. Hirsch, U.S. Geological Survey; Judith L. Meyer, University of Georgia; Larry A. Roesner, Colorado State University; Robert V. Thomann, Manhattan University (retired); and Robert C. Ward, Colorado State University.

Although the reviewers listed above have provided many constructive comments and suggestions, they were not asked to endorse the conclusions or recommendations, nor did they see the final draft of the report before its release. The review of this report was overseen by Frank H. Stillinger, Princeton University, and D. Peter Loucks, Cornell University. Appointed by the NRC, they were responsible for making certain that an independent examination of this report was carried out in accordance with institutional procedures and that all review comments were carefully considered. Responsibility for the final content of this report rests entirely with the authoring committee and the NRC.

KENNETH H. RECKHOW
Chair

Contents

EXECUTIVE SUMMARY, *1*
    TMDL Program Goals, 3
    Changes to the TMDL Process, 4
    Use of Science in the TMDL Program, 7
    Final Thoughts, 11

1   INTRODUCTION, *12*
    The Return to Ambient-Based Water Quality Management, 12
    National Research Council Study, 16
    Current TMDL Process and Report Organization, 19
    References, 21

2   CONCEPTUAL FOUNDATIONS FOR WATER QUALITY MANAGEMENT, *22*
    Ambient Water Quality Standards, 22
    Decision Uncertainty, 28
    Conclusions and Recommendations, 30
    References, 31

3   WATERBODY ASSESSMENT: LISTING AND DELISTING, *32*
    Adequate Ambient Monitoring and Assessment, 33
    Defining All Waters, 42
    Desirable Criteria, 44
    Listing and Delisting in a Data-Limited Environment, 50
    Data Evaluation for the Listing and Delisting Process, 56
    Use of Models in the Listing Process, 61
    References, 63

4 MODELING TO SUPPORT THE TMDL PROCESS, 68
   Model Selection Criteria, 69
   Uncertainty Analysis in Water Quality Models, 73
   Models for Biotic Response: A Critical Gap, 77
   Additional Model Selection Issues, 80
   References, 86

5 ADAPTIVE IMPLEMENTATION FOR IMPAIRED WATERS, 89
   Science and the TMDL Process, 89
   Review of Water Quality Standards, 90
   Adaptive Implementation Described, 94
   TMDL Implementation Challenges, 97
   References, 102

APPENDIXES

A List of Guest Presentations at the First Committee Meeting, 103

B Biographies of the Committee Members and NRC Staff, 105

Executive Summary

Over the last 30 years, water quality management in the United States has been driven by the control of point sources of pollution and the use of effluent-based water quality standards. Under this paradigm, the quality of the nation's lakes, rivers, reservoirs, groundwater, and coastal waters has generally improved as wastewater treatment plants and industrial dischargers (point sources) have responded to regulations promulgated under authority of the 1972 Clean Water Act. These regulations have required dischargers to comply with effluent-based standards for criteria pollutants, as specified in National Pollutant Discharge Elimination System (NPDES) permits issued by the states and approved by the U.S. Environmental Protection Agency (EPA). Although successful, the NPDES program has not achieved the nation's water quality goals of "fishable and swimmable" waters largely because discharges from other unregulated nonpoint sources of pollution have not been as successfully controlled. Today, pollutants such as nutrients and sediment, which are often associated with nonpoint sources and were not considered criteria pollutants in the Clean Water Act, are jeopardizing water quality, as are habitat destruction, changes in flow regimes, and introduction of exotic species. This array of challenges has shifted the focus of water quality management from effluent-based to ambient-based water quality standards.

This is the context in which EPA is obligated to implement the Total Maximum Daily Load (TMDL) program, the objective of which is attainment of ambient water quality standards through the control of both point and nonpoint sources of pollution. Although the TMDL program originated from Section 303d of the Clean Water Act, it was largely overlooked during the 1970s and 1980s as states focused on bringing point sources of pollution into compliance with NPDES permits. Citizen lawsuits during the 1980s forced EPA to develop guidance for the TMDL program, which is now considered to be pivotal in securing the nation's water quality goals.

Under TMDL regulations promulgated in 1992, EPA requires states to list waters that are not meeting water quality criteria set for specific designated uses. For each impaired water, the state must identify the amount by which point and nonpoint sources of pollution must be reduced in order for the waterbody to meet its stated water quality standards. Meeting these requirements, many of which have been imposed by court order or consent decree, has become the most pressing and significant regulatory water quality challenge for the states since passage of the Clean Water Act.

Given the most recent lists of impaired waters submitted to EPA, there are about 21,000 polluted river segments, lakes, and estuaries making up over 300,000 river and shore miles and 5 million lake acres. The number of TMDLs required for these impaired waters is greater than 40,000. Under the 1992 EPA guidance or the terms of lawsuit settlements, most states are required to meet an 8- to 13-year deadline for completion of TMDLs. Budget requirements for the program are staggering as well, with most states claiming that they do not have the personnel and financial resources necessary to assess the condition of their waters, to list waters on 303d, and to develop TMDLs. A March 2000 report of the General Accounting Office (GAO) highlighted the pervasive lack of data at the state level available to set water quality standards, to determine what waters are impaired, and to develop TMDLs.

Subsequent to the GAO report and following issuance by EPA of updated TMDL regulations, Congress requested that the National Research Council (NRC) assess the *scientific basis* of the TMDL program, including:

- the information required to identify sources of pollutant loadings and their respective contributions to water quality impairment,
- the information required to allocate reductions in pollutant loadings among sources,
- whether such information is available for use by the states and whether such information, if available, is reliable, and
- if such information is not available or is not reliable, what methodologies should be used to obtain such information.

Of concern to the nation's lawmakers was the paucity of data and information available to the states to comply with program requirements and meet water quality standards. Indeed, as the TMDL program proceeds, the best available science, especially with regard to nonpoint sources of pollution, will be needed for regulatory and nonregulatory actions to be equitable and effective. Report recommendations are targeted (1) at those issues where science can and should make a significant contribution and (2) at barriers (regulatory and otherwise) to the use of science in the TMDL program. Chapters 2, 3, and 4 discuss the information required to set water quality standards, to list waters as impaired, and to develop TMDLs (including the identification of pollution sources), while Chapter 5 discusses the role of science in allocating pollutant loading among sources. Chapters 3 and 4 go into considerable detail about the monitoring, modeling, and statistical analysis methods needed to collect data and convert it to information, and to assess and reduce uncertainty.

This report represents the consensus opinion of the eight-member NRC committee assembled to complete this task. The committee met three times during a three-month period and heard the testimony of over 40 interested organizations and stakeholder groups. The NRC committee feels that the data and science have progressed sufficiently over the past 35 years to support the nation's return to ambient-based water quality management. Given reasonable expectations for data availability and the inevitable limits on our conceptual understanding of complex systems, statements about the science behind water quality management must be made with acknowledgment of uncertainties. The committee has concluded that there are creative ways to accommodate this uncertainty while moving forward in addressing the nation's water quality challenges. These broad conclusions are elaborated upon below.

TMDL PROGRAM GOALS

The TMDL program should focus first and foremost on improving the condition of waterbodies as measured by attainment of designated uses. Work on meeting the strict time demands within the budget constraints cited by most states has focused on administrative outcomes as measures of success for the TMDL program. However, the success of the nation's premier water quality program should not be measured by the number of TMDL plans completed and approved, nor by the number of NPDES permits issued or cost share dollars spent. Success is achieved when the condition of a waterbody supports its designated use. Adequate monitoring and assessment must be used to improve the listing of impaired waterbodies and to characterize the effectiveness of the actions taken to meet the designated use.

The program should encompass all stressors, both pollutants and pollution, that determine the condition of the waterbody[1]. Proposed regulations may limit the applicability of the program to only those water quality problems caused by chemical and physical pollutants. Given their demonstrated effectiveness, activities that can overcome the effects of "pollution" and bring about waterbody restoration—such as habitat restoration and channel modification—should not be excluded from consideration during TMDL plan implementation.

Scientific uncertainty is a reality within all water quality programs, including the TMDL program, that cannot be entirely eliminated. The states and EPA should move forward with decision-making and implementation of the TMDL program in the face of this uncertainty while making substantial efforts to reduce uncertainty. Securing designated uses is limited not only by a focus on administrative rather than water quality outcomes in the TMDL process, but also by unreasonable expectations for predictive certainty among regulators, affected sources, and stakeholders.

CHANGES TO THE TMDL PROCESS

This report focuses on how scientific data and information should be used within the TMDL program. Science plays a crucial role in the standards-setting process, in the decision to add waters to the 303d list, in the development of the TMDL plan, and in the allocation of pollutant loads among various sources (although its importance relative to the role of policy decisions varies). The committee finds that although the state of the science is sufficient to develop TMDLs to meet ambient water quality goals in many situations, programmatic issues substantially hinder the use of the best available science. Thus, the following changes in the TMDL process are recommended, with an understanding that without such changes, the TMDL program will be unable to incorporate and improve upon the best available scientific information.

States should develop appropriate use designations for waterbodies in advance of assessment and refine these use designations prior to TMDL development. Clean Water Act goals of fishable and swimmable waters are too broad to be operational as statements of designated uses.

---

[1] This refers to the legal definitions of "pollutant" and "pollution," which are given in Box 1-1 of Chapter 1.

Thus, there should be greater stratification of designated uses at the state level (such as primary and secondary contact recreation). The appropriate designated use may not be the use that would be realized in the water's predisturbance condition. Sufficient science and examples exist for all states to inject this level of detail into their water quality standards. To ensure that designated uses are appropriate, use attainability analysis should be considered for all waterbodies before a TMDL is developed.

EPA should approve the use of both a preliminary list and an action list instead of one 303d list. Many waters now on state 303d lists were placed there without the benefit of adequate water quality standards, data, or waterbody assessment. These potentially erroneous listings contribute to a very large backlog of TMDL segments and foster the perception of a problem that is larger than it may actually be. States should be allowed to move those waters for which there is a lack of adequate water quality standards or data and analysis from the 303d list back to a preliminary list, as shown in Figure ES-1. This would provide the assurance that listed waters are indeed legitimate and merit the resources required to complete a TMDL. If no legal mechanism exists to bring this about, one should be created by Congress. The data requirements and other criteria that should be used to differentiate the preliminary list from the action list are discussed in the report. No waterbody should remain on the preliminary list for more than one rotating basin cycle.

TMDL plans should employ adaptive implementation. As shown in Figure ES-2, adaptive implementation is a cyclical process in which TMDL plans are periodically assessed for their achievement of water quality standards including designated uses. If the implementation of the TMDL plan is not achieving attainment of the designated use, scientific data and information should be used to revise the plan. Adaptive implementation is needed to ensure that the TMDL program is not halted because of a lack of data and information, but rather progresses while better data are collected and analyzed with the intent of improving upon initial TMDL plans. Congress and EPA need to address the policy barriers that inhibit adoption of an adaptive implementation approach to the TMDL program, including the issues of future growth, the equitable distribution of cost and responsibility among sources of pollution, and EPA oversight.

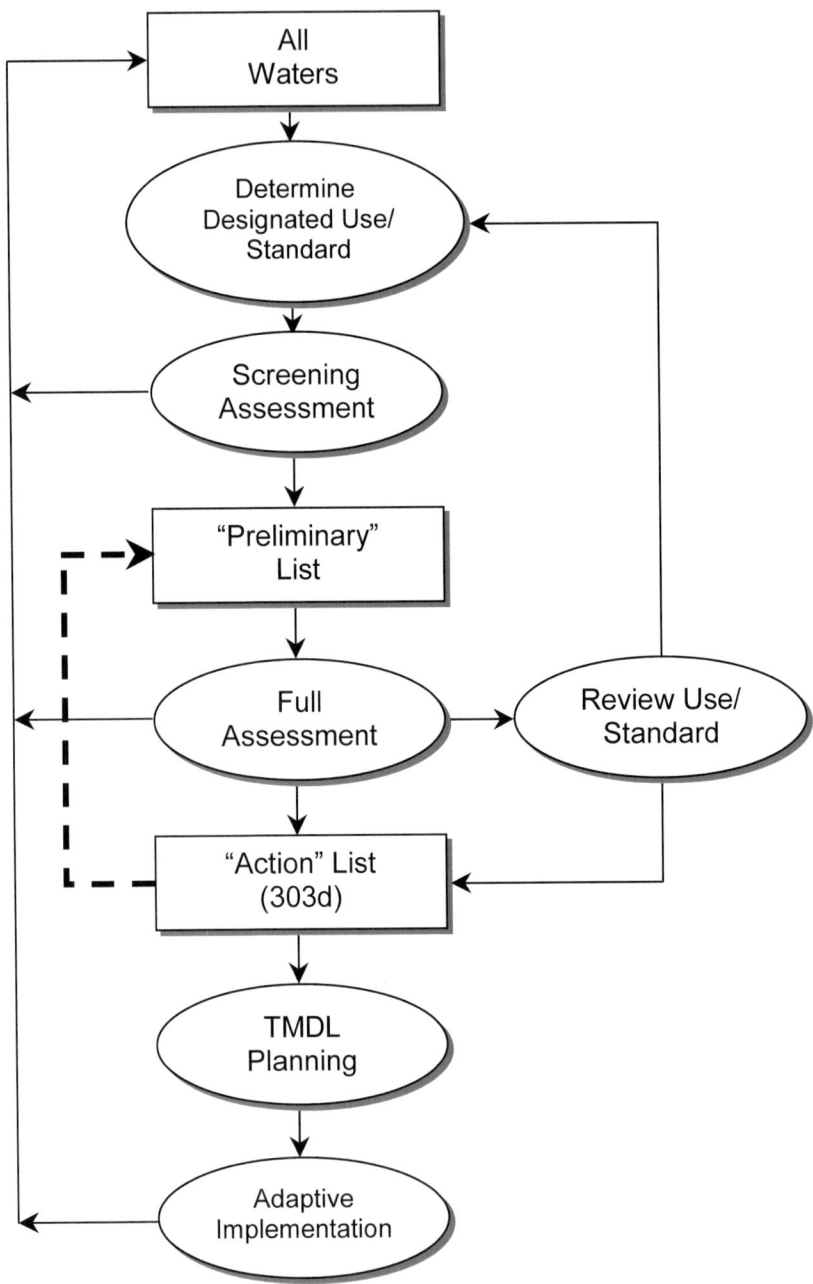

FIGURE ES-1 Framework for water quality management.

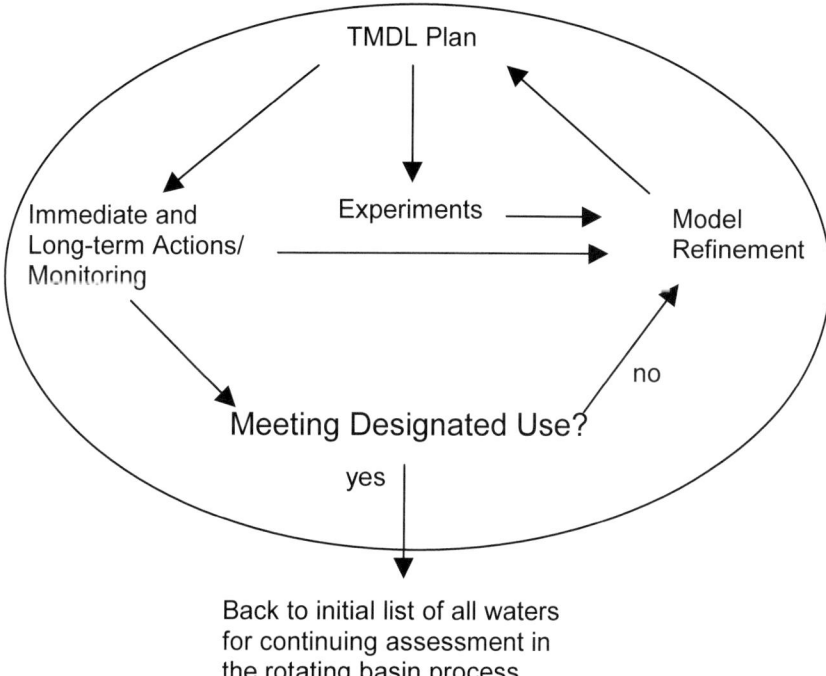

FIGURE ES-2 Adaptive implementation flowchart.

USE OF SCIENCE IN THE TMDL PROGRAM

This report suggests changes in the data used and analytical methods employed that will support the revisions to the TMDL process recommended above. The following sections highlight the use of science in the TMDL program steps as illustrated in Figure ES-1. Additional recommendations about the scientific basis of the program not included in this executive summary are found throughout the report.

Water Quality Standards

The TMDL process is primarily a measurement process and as such is significantly impacted by the setting of water quality standards. Water quality standards consist of two parts: a specific desired use appropriate to the waterbody, termed a *designated use,* and a *criterion* that can be measured to establish whether the designated use is being achieved.

The criterion used to measure whether the condition of a waterbody supports its designated use can be positioned at different points along the causal chain connecting stressors (such as land use activities) to biological responses in a waterbody. Positioning the criterion involves a trade-off between forecast error for the stressor–criterion relationship and the adequacy of the criterion as a measure (surrogate) for the designated use. Model results that forecast the impact of the stressor on the criterion are likely to be more uncertain as the criterion is positioned farther from the stressor and closer to the designated use. On the other hand, positioning the criterion closer to the stressor and farther from the designated use is likely to mean that the criterion is a poorer measure or surrogate for the designated use.

Biological criteria should be used in conjunction with physical and chemical criteria to determine whether a waterbody is meeting its designated use. In general, biological criteria are more closely related to the designated uses of waterbodies than are physical or chemical measurements. However, guiding management actions to achieve water quality goals based on biological criteria also depends on appropriate modeling efforts.

All chemical criteria and some biological criteria should be defined in terms of magnitude, frequency, and duration. The frequency component should be expressed in terms of a number of allowed excursions in a specified period. Establishing these three dimensions of the criterion is crucial for successfully developing water quality standards and subsequently TMDLs.

Water quality standards must be measurable by reasonably obtainable monitoring data. In many states, there is a fundamental discrepancy between the criteria that have been chosen to determine whether a waterbody is achieving its designated use and the frequency with which water quality data are collected. This report gives examples of this phenomenon and makes suggestions for improvement.

Waterbody Assessment and Listing

Ambient monitoring and assessment programs should form the basis for determining whether waters are placed on the preliminary list or the action list.

Executive Summary 9

EPA needs to develop a uniform, consistent approach to ambient monitoring and data collection across the states. The rotating basin approach used by several states is an excellent example of a framework than can be used to conduct waterbody assessments of varying levels of complexity, for example to support 305b reports, to place impaired waters on a preliminary list or action list, and to develop TMDLs. **In that regard, EPA should set the TMDL calendar in concert with each state's rotating basin program.**

Evidence suggests that limited budgets are preventing the states from monitoring for a full suite of indicators to assess the condition of their waters and from embracing a rotating basin approach to water quality management. Currently, EPA is assessing the sufficiency of state resources to develop and implement TMDLs. Depending on the results of that assessment, Congress might consider aiding the states, for example through matching grants to improve data collection and analysis.

Evaluated data and evidence of violation of narrative standards should not be exclusively used for placement of a waterbody on the action list, but is useful for placement on the preliminary list. EPA should develop guidance to help states translate narrative standards to numeric criteria for the purposes of 303d listing and TMDL calculation and implementation.

EPA should endorse statistical approaches to defining all waters, proper monitoring design, data analysis, and impairment assessment. For chemical parameters, these statistical approaches might include the binomial hypothesis test or other methods that can be more effective than the raw score approach in making use of the data collected to determine water quality impairment. For biological parameters, they might focus on improvement of sampling designs, more careful identification of the components of biology used as indicators, and analytical procedures that explore biological data as well as integrate biological information with other relevant data.

TMDL Development

The scientific basis of the latter half of the TMDL process revolves around a wide variety of models of varying complexity that are used to relate waterbody conditions to different land uses and other factors. Models are a required element of developing TMDLs because water quality standards are probabilistic in nature. However, although models can aid in the decision-making process, they do not eliminate the need for informed decision-making.

Uncertainty must be explicitly acknowledged both in the models selected to develop TMDLs and in the results generated by those models. Prediction uncertainty must be estimated in a rigorous way, models must be selected and rejected on the basis of a prediction error criterion, and guidance/software needs to be developed to support uncertainty analysis.

The TMDL program currently accounts for the uncertainty embedded in the modeling exercise by applying a margin of safety (MOS); EPA should end the practice of arbitrary selection of the MOS and instead require uncertainty analysis as the basis for MOS determination. Because reduction of the MOS can potentially lead to a significant reduction in TMDL implementation cost, EPA should place a high priority on selecting and developing TMDL models with minimal forecast error.

EPA should selectively target some postimplementation TMDL compliance monitoring for verification data collection so that model prediction error can be assessed. TMDL model choice is currently hampered by the fact that relatively few models have undergone thorough uncertainty analysis. Postimplementation monitoring at selected sites can yield valuable data sets to assess the ability of models to reliably forecast response.

EPA should promote the development of models that can more effectively link environmental stressors (and control actions) to biological responses. A first step will be the development of conceptual models that account for known system dynamics. Eventually, these should be strengthened with both mechanistic and empirical models, although empirical models are more likely to fill short-term needs. Such models are needed to promote the wider use of biocriteria.

Monitoring and data collection programs need to be coordinated with anticipated water quality and TMDL modeling requirements. For many parameters, there are insufficient data to have confidence in the results generated by some of the complex models used in practice today. Thus, EPA should not advocate detailed mechanistic models for TMDL development in data-poor situations. Either simpler, possibly judgmental, models should be used or, preferably, data needs should be anticipated so that these situations are avoided.

In order to carry out adaptive implementation, EPA needs to foster the use of strategies that combine monitoring and modeling and expedite TMDL development. This should involve the use of Bayesian techniques that can combine different types of information. Although the modeling framework proposed in this report calls for improvements in models, there are existing models that can be applied rapidly and effectively within an adaptive implementation framework.

FINAL THOUGHTS

Through the adoption and use of the preliminary list/action list approach, adequate monitoring and assessment approaches, sound selection of appropriate models, and adaptive implementation described in this report, the TMDL program will be capable of utilizing the best available scientific information. It is worth noting that the success of these approaches is directly related to the provision of adequate personnel and financial resources for data collection, management, and interpretation and for the development of sufficiently detailed and stratified water quality standards.

1
Introduction

THE RETURN TO AMBIENT-BASED WATER QUALITY MANAGEMENT

The Federal Water Pollution Control Act Amendments of 1972 (PL 92-500), as supplemented by the Clean Water Act (CWA) of 1977 and the Water Quality Act of 1987, are the foundation for protecting the nation's water resources. Precursors to the Water Quality Act go back to the Rivers and Harbors Appropriations Act of 1899, often referred to as the Refuse Act, and the Water Pollution Control Acts of 1948 and 1965 (Rodgers, 1994). An important impetus for earlier water quality legislation was protection of public health. Over time, this purpose was supplemented by aesthetic and recreational purposes (fishable and swimmable) and then by the goal of restoring and maintaining the "chemical, physical, and biological integrity of the Nation's waters" (Section 101a of PL 92-500).

In practice, each of these general purposes must be restated in operational and measurable terms as *ambient* water quality standards, which are established by the states and are subject to federal approval. Section 303d of the CWA makes it a responsibility of the states to assess whether ambient standards are being achieved for individual waterbodies. If ambient standards are not being met, a water quality management program to achieve those standards is anticipated.

The data and analytical requirements for determining both the causes of a failure to meet ambient standards and the solutions to such problems have challenged water quality analysts for over half a century. Prior to the 1972 Water Pollution Control Act Amendments, states were expected to identify pollutant sources that were resulting in violations of ambient water quality standards. Once the sources of the problem were carefully identified, controls on polluting activities would be put in place. However, in even modestly complex watersheds, multiple sources of pollutants made it difficult to unambiguously determine which sources were responsible for the standard violation. One source might insist that the cause of the problem was the discharge from others, or at least that its own contribution to the problem was not as significant as the contributions of others. Neither the available monitoring data nor the analytical methods available at the time allowed the states to defensibly mandate differential load reduction requirements (Houck, 1999).

The 1972 amendments recognized this analytical dilemma and shifted the focus of water quality management away from ambient standards. Instead, all dischargers of certain pollutants were expected to limit their discharges by meeting nationally established *effluent standards*. Effluent standards are specified in National Pollution Discharge Elimination System (NPDES) permits, issued by the states to certain pollutant sources and approved by the U.S. Environmental Protection Agency (EPA). Effluent standards were set at a national level based on available technologies for wastewater treatment appropriate to different industry groups (although in certain waterbodies effluent standards more stringent than the technology-based requirement have been required to meet local water quality goals). The shift to effluent standards eliminated the need to link required reductions at particular sources with the ambient condition of a waterbody. Instead, each regulated source was simply required to meet the effluent standard in its wastewater. In the intervening period since passage of PL 92-500, pollutants discharged by industry and municipal treatment plants have declined, and the ambient quality of many of the nation's lakes, rivers, reservoirs, groundwater, and coastal waters has improved.

There were consequences that followed the embracing of effluent-based standards instead of ambient-based standards. First, efforts to measure and communicate water quality accomplishments were often described in terms of compliance with wastewater permit conditions rather than the condition of the waters. Second, effluent standards could only apply to so-called point sources rather than to all sources of a pollutant or other forms of pollution (Box 1-1). Pollutants from nonpoint sources (derived from diffuse and hard-to-monitor origins such as land-disturbing agricultural, silvicultural, and construction activities) largely escaped oversight. Third, attention to chemical pollutants measured in discharge water came to dominate water quality policy, and the physical and biological determinants of the ambient condition of a waterbody were less frequently considered. A *pollutant* is defined as a substance added by humans or human activities. In many cases, the condition of a

> **BOX 1-1**
> **Pollution vs. Pollutant**
>
> **Clean Water Act Section 502(6).** The term "pollutant" means dredged spoil, solid waste, incinerator residue, biological materials, radioactive materials, heat, wrecked or discarded equipment, rock, salt, cellar dirt, and industrial, municipal, and agricultural waste discharged into water. This term does not mean (A) "sewage from vessels" within the meaning of section 312 of this Act; or (B) water, gas, or the materials which are injected into a well to facilitate production of oil or gas, or water derived in association with oil or gas production and disposed of in a well, if the well used either to facilitate production or for disposal purposes is approved by authority of the State in which the well is located, and if such State determines that such injection or disposal will not result in the degradation of ground or surface water resources.
>
> **Clean Water Act Section 502(19).** The term "pollution" means the manmade or man-induced alteration of chemical, physical, biological, and radiological integrity of water.
>
> In the Clean Water Act, pollution includes pollutants (as described above) as well as other stressors such as habitat destruction, hydrologic modification, etc.

waterbody depends on more than the loads of particular pollutants from sources required to meet effluent standards. For example, changes in the hydrologic regime associated with development activities can destabilize streambanks, increase loads of sediment and nutrients, or eliminate key species or otherwise change the aquatic ecosystem. As shown in Box 1-1, biological, hydrologic, and physical changes to a waterbody that do not fit the definition of pollutant were encompassed in the 1987 act's definition of *pollution.*

Present-day implementation of Section 303d of the Clean Water Act returns to the pre-1972 focus on ambient water quality standards, even though there are still requirements for meeting effluent standards. Section 303d requires states to identify waters not meeting ambient water quality standards, define the pollutants and the sources responsible for the degradation of each listed water, establish Total Maximum Daily Loads (TMDLs) necessary to secure those standards, and allocate responsibility to sources for reducing their pollutant releases. Therefore, for each impaired waterbody, the state must identify the amount by which both point and nonpoint source pollutants would need to be reduced in order for the waterbody to meet ambient water quality standards. Other alterations that do not fit the pollutant definition such as changes of habitat, flow alterations, channelization, and modification or loss of riparian habitat may need to be considered as a reason for not meeting standards. If TMDL language is strictly interpreted, however, these causes may fall outside the TMDL program.

Although Section 303d has been in place since the early 1970s, activity to comply with it was limited until the last decade. States were slow to submit inventories of impaired waters, and measures of water quality program success were often simply documentation of point source permit issuance and compliance. Few TMDLs were prepared, and they often did not incorporate both point and nonpoint source discharge controls (Houck, 1999). Action to meet Section 303d requirements accelerated in the 1990s primarily because of a series of citizen lawsuits against EPA. By 1992, EPA revised the TMDL regulations to require submission of states' lists of impaired water bodies every two years.

EPA estimates that from 3,800 to 4,000 TMDLs will need to be completed per year to meet the 8- to 13-year deadlines currently imposed on the process. From 1,000 to 1,800 would have to be completed per year to meet consent decree deadlines, while another 1,800 to 2,200 per year need to be resolved through settlement agreements. States have identified about 21,000 impaired river segments, lakes, and estuaries encompassing more than 300,000 river and shore miles and 5 million lake acres (Brady, 2001). Excess sediments, nutrients, and pathogens are leading reasons for listing according to state reports submitted to EPA. Federal, state, and local governments, regulated and potentially regulated communities, and concerned citizens throughout the nation claim that they face unrealistic deadlines and must use analytical and decision-making procedures that are largely untested. Proposed revisions to the TMDL regulations were submitted in 1999, with a final rule issued July 13, 2000. However, faced with expressions of concern about the practicality of the program, a congressional rider prohibited EPA from implementing the new rule until October 2001. As a result, the TMDL program continues under 1992 regulations and, in some cases, consent decrees.

The 303d focus on ambient water quality standards has returned the nation to a water quality program that was not considered implementable 35 years ago when there was a paucity of data and analytical tools for determining causes of impairment and assigning responsibility to various sources. Determining the pollutant load from a regulated point source is a relatively straightforward task, although isolating its effect in a complex waterbody remains a technical challenge. Such technical uncertainties in relating stresses on the waterbody to impairment are compounded when nonpoint sources of pollutants and other forms of pollution are considered. Having returned the focus to ambient water quality conditions, are we better positioned today than we were years ago? Do we have more and better data and analytical methods? Do we have a better understanding of watershed events and processes responsible for water quality violations? These are the science questions facing the nation as we implement Section 303d of the Clean Water Act.

NATIONAL RESEARCH COUNCIL STUDY

Despite recent progress, the demands of the TMDL program weigh heavily on the limited resources of EPA and the states. The TMDL process requires high-quality data and sophisticated tools to analyze those data. States have reported having insufficient funds, inadequate monitoring programs, and limited staff to collect and analyze such data (GAO, 2000). According to the General Accounting Office (GAO), only six states have enough data to fully assess the condition of their waterbodies, while only 18 have enough data to place their waterbodies on the list of impaired waters (303d list). Forty states had sufficient high-quality data to determine TMDLs for waterbodies impaired primarily by point sources such as municipal sewage treatment plants, and 29 had sufficient high-quality data to implement these TMDLs. When states were asked about waterbodies impaired primarily by nonpoint sources, however, only three claimed to have sufficient data.

The GAO report outlined several critical issues for consideration by the states and EPA. Beyond questions of additional funding for data collection and staff, the states need assistance using watershed models; many reported being unclear where to go for such assistance. There appears to be no formalized process to capitalize on lessons learned, to transfer technology, and to share knowledge. Aside from the reported

Conceptual Foundations for Water Quality Management

lack of data to comply with the TMDL regulations, when data are available, they are often not the type needed for source identification and TMDL analyses.

Subsequent to the GAO report, Congress requested that the National Research Council (NRC) analyze on a broad scale *the scientific basis of the TMDL program*. The NRC was asked to evaluate:

- the information required to identify sources of pollutant loadings and their respective contributions to water quality impairment,
- the information required to allocate reductions in pollutant loadings among sources,
- whether such information is available for use by the states and whether such information, if available, is reliable, and
- if such information is not available or is not reliable, what methodologies should be used to obtain such information.

While the GAO report was about data, the NRC was charged to focus on *reliable information* for making decisions. In presentations made to the NRC committee, the terms "data" and "information" often were used as synonyms, but data are not the same as information. Unanalyzed data do not constitute information. Data must be interpreted for their meaning through the filter of analytical techniques, and the result of such data analysis is information that can support decision-making. Knowing what data are needed and turning those data into information constitutes, in large part, the science behind a water quality management program. The techniques for transforming data into information include statistical inference methods, simulation modeling of complex systems, and, at times, simply the application of the best professional judgment of the analyst. In all these processes there will always be some uncertainty (and thus some "unreliability") about whether the resulting information accurately characterizes the water quality problem and the effectiveness of the solutions. Because uncertainty cannot be eliminated, determining whether the information generated from data analysis is reliable is a value judgment. Individuals and groups will have different opinions about whether and how to proceed with water quality management given a certain level of uncertainty.

To organize its deliberations, the committee considered the role of science at each step of the TMDL process, from the initial defining of all waters to the implementation of actions to control pollution; the report is structured around this organization. Report recommendations are targeted (1) at those issues where science can and should make a significant contribution and (2) at barriers (regulatory and otherwise) to the use of science in the TMDL program. Because of this broad scope, the content of the report extends beyond the confines of the charge in the bulleted items above. Chapters 2, 3, and 4 discuss the information (as defined above) required to set water quality standards, to list waters as impaired, and to develop TMDLs (including the identification of pollution sources); Chapter 5 comments on the role of science in allocating pollutant loading among sources. Because GAO (2000) already documents a widespread lack of data and information at the state level and because availability of information varies significantly from state to state, the committee did not devote substantial time to determining availability. As mentioned above, whether the information is reliable depends on the degree of uncertainty decision-makers are willing to accept when making regulatory or spending choices—a decidedly nonscientific matter. Chapters 3 and 4 describe in detail the monitoring, modeling, and statistical analysis methods needed to collect data and convert it to information, and to assess and reduce uncertainty. Chapter 5 describes an approach for making decisions in the face of uncertainty.

This report represents the culmination of three meetings over three months, including a two-day public session in which 30 presentations from a wide variety of stakeholders were made (see Appendix B). Given the information gathered during the study period and the collective experience of its members, the committee feels that the data and science have progressed sufficiently over the past 35 years to support the nation's return to ambient-based water quality management. In addition, the need for this approach is made apparent by the inability of a large percentage of the nation's water to meet water quality standards using point source controls alone. Given reasonable expectations for data availability and inevitable limits on our conceptual understanding of complex systems, statements about the science behind water quality management must be made with acknowledgment of uncertainties. Finally, the committee has concluded that there are creative ways to accommodate this uncertainty while moving forward in addressing the nation's water quality challenges. These broad conclusions are elaborated upon throughout this report.

CURRENT TMDL PROCESS AND REPORT ORGANIZATION

Section 303d requires that states identify waters that are not attaining ambient water quality standards (i.e., are impaired). (Although new rules are pending, at the request of Congress, this report focuses on the 1992 regulations that govern the current program.) States must then establish a priority ranking for such waters, taking into account the severity of the impairment and the uses to be made of such waters. For impaired waters, the states must establish TMDLs for pollutants necessary to secure applicable water quality standards. The CWA further requires that once water quality standards are attained they must be maintained.

Figure 1-1 depicts the basic steps in the TMDL process. These steps are described briefly below and are considered in greater detail throughout the report. At the beginning of the process are all waterbodies for the state and the development of water quality standards for each waterbody. Water quality standards are established outside the TMDL process and include designated uses for a waterbody and measurable water quality

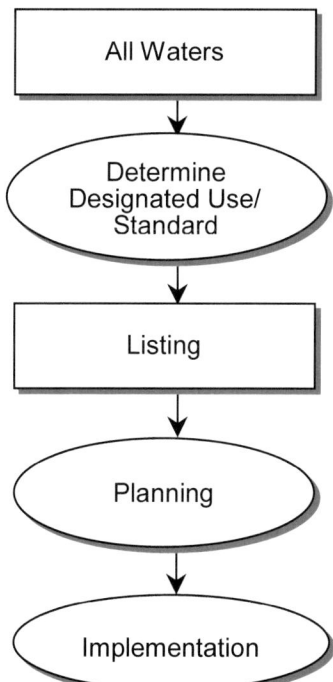

FIGURE 1-1 Conceptualized steps of the TMDL process.

criteria designed to assure that each designated use is being achieved.

Because water quality standards are the foundation on which the entire TMDL program rests, more detailed discussion of standard setting is provided in Chapters 2 and 3.

The next step in the process is the listing of *impaired* waterbodies if evaluation of available data suggests that certain waterbodies are not meeting standards. According to Section 303d, all impaired waterbodies must be listed by the states or responsible agencies and submitted to EPA every two years. In addition, the states should provide priority ranking for the waterbodies on the 303d list. Following its submission, EPA must either approve or disapprove the list. Listing of a waterbody initiates a costly planning process and may lead to added costs to implement pollutant controls by point and nonpoint sources. The NRC committee heard testimony that many waterbodies have been listed based on limited or completely absent data and poorly conceived analytical techniques for data evaluation. Chapter 3 reviews the listing process and makes recommendations that will improve the reliability of the listing decision.

Once an impaired waterbody is listed, a planning step ensues. Section 303d specifies that those waters impaired by pollutants should undergo calculation of a TMDL. The term TMDL has essentially two meanings (EPA, 1991):

- The TMDL process is used for implementing state water quality standards—that is, it is a planning process that will lead to the goal of meeting the water quality standards.
- The TMDL is a numerical quantity determining the present and near future maximum load of pollutants from point and nonpoint sources as well as from background sources, to receiving waterbodies that will not violate the state water quality standards with an adequate margin of safety. The permissible load is then allocated by the state agency among point and nonpoint sources.

The calculation described above requires data collection and various forms of modeling in order to identify sources of pollution and background conditions, calculate the maximum load that will meet water quality standards with a margin of safety, and make allocations of responsibility for load reduction to point and nonpoint sources. Chapter 4 reviews modeling capability, data needs for model implementation, and the appropriate role of modeling in the TMDL planning process.

The last step in the process is implementation of the TMDL and the delisting of the waterbody. Implementation is the process of putting the actions envisioned in the TMDL plan in place. Such actions could include limitations on point sources beyond technology-based effluent standards. Also, using best management practices for nonpoint sources, as well as addressing pollution problems, might be part of implementation, although these actions are not required by Section 303d.[2] The results of implementation actions need to be assessed before a waterbody can be removed from the list. Monitoring in this phase is necessary to measure the success (or failure) of the plan. Chapter 5 discusses postimplementation monitoring and a strategy for assuring that the best available science is used in the TMDL implementation phase. When the monitoring proves that the implementation is successful (i.e., the water quality standards are met), the waterbody can be delisted.

REFERENCES

Brady, D. 2001. Chief of the Watershed Branch in the Assessment and Watershed Protection Division in the EPA Office of Wetlands, Oceans and Watersheds. Presentation to the NRC Committee. January 25, 2001.
Environmental Protection Agency (EPA). 1991. Guidance for Water Quality-based Decisions: The TMDL Process. Washington, DC: EPA Assessment and Watershed Protection Division.
General Accounting Office (GAO). 2000. Water Quality - Key EPA and State Decisions Limited by Inconsistent and Incomplete Data. GAO/RCED-00-54. Washington, DC: GAO.
Houck, O. A. 1999. The Clean Water Act TMDL Program: Law, Policy, and Implementation. Washington, DC: Environmental Law Institute.
Rodgers, W. H., Jr. 1994. Environmental Law, Second edition. St. Paul, MN: West Publishing Co.

---

[2] Whether nonpoint source controls are required as part of the TMDL program is the source of much of the debate, especially with regard to the 2000 regulations that are now on hold. Under the current (1992) regulations, 303d is a planning exercise only. Implementation must be by some other provisions of the CWA or other programs. Also, states differ in their ability to enforce use of certain best management practices.

2
Conceptual Foundations for Water Quality Management

This chapter describes the analytical and related policy challenges of implementing an ambient-focused water quality management program, of which the Total Maximum Daily Load (TMDL) program is an example[3]. The goal of an ambient water quality management program is to measure the condition of a waterbody and then determine whether that waterbody is meeting water quality standards. By definition, this process is dependent on the setting of appropriate water quality standards. Although realistic standard setting must account for watershed (hydrologic, ecological, and land use) conditions, the corresponding need to make policy decisions in setting standards must also be recognized. In addition, ambient-based water quality management requires decision-making under uncertainty because the possibility for making assessment errors is always present. Properly executed statistical procedures can identify the magnitude and direction of the possible errors so that knowledge can be incorporated into the decisions made. In addition to uncertainties inherent in measuring the attainment of water quality standards, there are uncertainties in results from models used to determine sources of pollution, to allocate pollutant loads, and to predict the effectiveness of implementation actions on attainment of a standard. As part of the information needed in the TMDL program, this uncertainty must be understood and addressed as implementation decisions are made.

AMBIENT WATER QUALITY STANDARDS

Unlike an effluent standard, an ambient water quality standard ap-

---

[3] Although this discussion refers to the TMDL program, it is not meant to be a description of that program.

plies to a specific spatial area—a defined waterbody—and is expected to be met over all areas of that waterbody. Thus, identifying the waterbody of interest, whether a lake, a stream segment, or areas of an estuary, is a first step in setting water quality standards. Waterbodies vary greatly in size—for example, from a small area such as a mixing zone below a point source discharge on a river to an estuary formed by a major river discharge.

Water quality standards themselves consist of two parts: a specific desired use appropriate to the waterbody, termed a *designated use*, and a *criterion* that can be measured to establish whether the designated use is being achieved. Barriers to achieving the designated use are the presence of pollutants and hydrologic and geomorphic alterations to the waterbody or watershed.

Appropriate Designated Uses

A designated use describes the goal of the water quality standard. For example, a designated use of human contact recreation should protect humans from exposure to microbial pathogens while swimming, wading, or boating. Other uses include those designed to protect humans and wildlife from consuming harmful substances in water, fish, and shellfish. Aquatic life uses are intended to promote the protection and propagation of fish, shellfish, and wildlife resources.

A designated use is stated in a written, qualitative form, but the description should be as specific as possible. Thus, more detail than "recreational support" or "aquatic life support" is needed. The general "fishable" and "swimmable" goals of the Clean Water Act constitute the beginning, rather than the end, of appropriate use designation. For example, a sufficiently detailed designated use might distinguish between beach use, primary water contact recreation, and secondary water contact recreation[4]. Similarly, rather than stating that the waterbody needs to be "fishable," the designated use would ideally describe whether the waterbody is expected to support a desired fish population (e.g., salmon, trout,

---

[4] These uses are defined differently from state to state. In Ohio, primary contact recreation includes full body immersion activities such as swimming, canoeing, and boating. Such streams or rivers must have a depth of at least 1 meter. Secondary contact recreation includes activities such as wading, but where full body immersion is not practical because of depth limitations. The fecal bacteria criteria are less stringent for secondary contact recreation than for primary contact recreation.

or bass) and the relative invertebrate or other biological communities necessary to support that population. Although small headwater streams may have aesthetic values, they may not have the ability to support extensive recreational uses themselves (i. e., be "fishable" or "swimmable"). However, their condition may have an influence on the ability of a downstream area to achieve a particular designated use. In this case, the designated use for the smaller waterbody may be defined in terms of the achievement of the designated use of the larger downstream waterbody (as illustrated in the discussion of criteria below).

In many areas of the United States, human activities have radically altered the landscape and aquatic ecosystems, such that an appropriate designated use may not necessarily be the aquatic life condition that was present in a watershed's predisturbance condition, which may be unattainable. For example, a reproducing trout fishery in downtown Washington, D.C., may be desired, but may not be attainable because of the development history of the area or the altered hydrologic regime of the waterbody. Similarly, designating an area near the outfall of a sewage treatment plant for shellfish harvesting may be desired, but health considerations would designate it as a restricted shellfish harvest water. Furthermore, there may be a conscious decision to establish a designated use that would *not* have existed in the predisturbance condition. For example, construction of a lake for a warm water fishery is a use possible only as a result of human intervention.

Appropriate use designation for a state's waterbodies is a policy decision that can be informed by technical analysis. However, a final selection will reflect a social consensus made in consideration of the current condition of the watershed, its predisturbance condition, the advantages derived from a certain designated use, and the costs of achieving the designated use. Ideally, a statewide water quality management program should establish a detailed gradient of use designations for waterbodies. Box 2-1 describes the multiple tiers of designated uses developed for waters in Ohio.

Defining a Criterion

A water quality standard includes a criterion representing the condition of the waterbody that supports the designated use. Thus, the designated use is a description of a desired endpoint for the waterbody, and the criterion is a measurable indicator that is a surrogate for use attainment.

BOX 2-1
Appropriate Designated Uses: The Ohio Example

An approach to setting appropriately stratified or tiered designated uses for a state's waterbodies has been developed in Ohio. The state recognized early on that a stratified set of use designations for aquatic life, recreation, and water supply was needed to accurately reflect the potential quality of various waterbodies and to guide cost-effective expenditures for pollution controls and other restoration activities. In lieu of general use, more detailed designated uses were developed that reflect the "potential" of the aquatic ecosystem and account for the historical influence of broad-scale socioeconomic activities. Individual waterbodies are assigned the appropriate designated use based on a use attainability analysis (UAA) process that relies heavily on site-specific information about the waterbody. The information used in this process results from the systematic monitoring of waters via a rotating basin approach in which biological, chemical, and physical data are collected and analyzed. Aquatic life uses are based primarily on the biological criteria and physical habitat assessments that are calibrated with regard to the important regional and watershed-specific variables that determine the potentially sustainable aquatic assemblage. Recreational uses are designated based on the size of the waterbody, reflecting the ability of humans to use the water for swimming, boating, fishing, or wading.

The system of tiered aquatic life and recreational uses in the Ohio water quality standards was established in 1978, well before biological criteria were adopted for use (May 1990). Two newly proposed uses are now under study: one for urban streams, which would require a site-specific UAA, and one for primary headwater streams (<1 sq. mi. drainage area), which are outside of the practical resolution of the present biological criteria. (A readily accessible and detailed example of such designated uses for Ohio can be found at http://www.epa.state.oh.us/dsw/rules/3745-1.html.

The criterion may be positioned at *any* point in the causal chain of squares shown in Figure 2-1. Criteria in squares 2 and 3 are possible measures of ambient water quality condition. Square 2 includes measures of a water quality parameter such as dissolved oxygen (DO), pH, nitrogen concentration, suspended sediment, or temperature. Criteria closer to the designated use (e.g., square 3) include measures such as the condition of the algal community (chlorophyll *a*), a comprehensive index

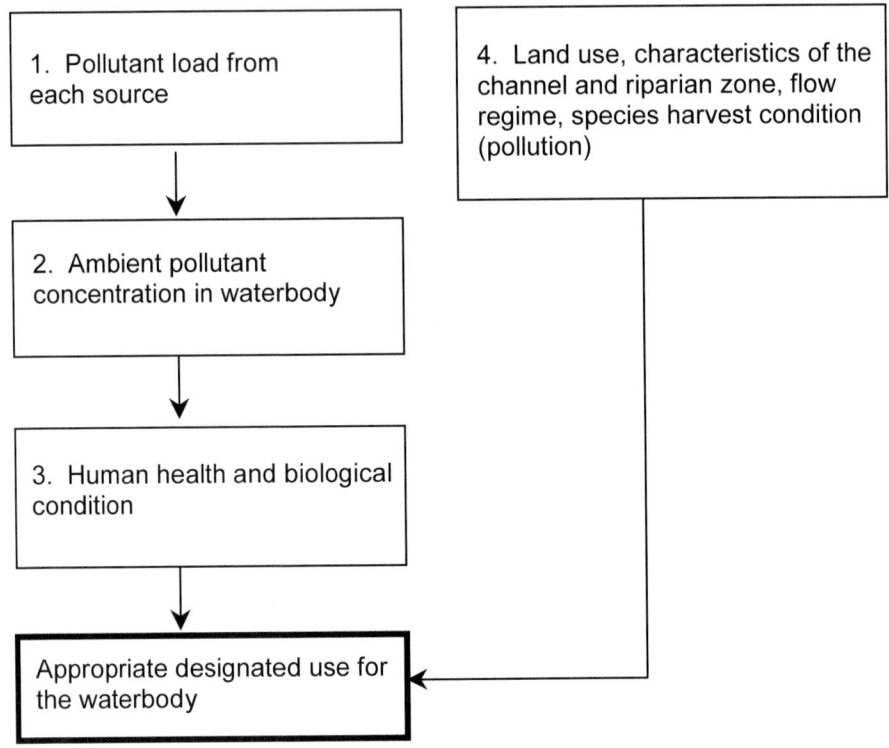

FIGURE 2-1 Types of water quality criteria and their position relative to designated uses.

measure of the biological community as a whole, or a measure of contaminant concentration in fish tissue. In square 1, where the criterion is farther from the designated use, are measures of the pollutant discharge from a treatment plant (e.g., biological oxygen demand, NH_3, pathogens, suspended sediments) or the amount of a pollutant entering the edge of a stream from runoff. A criterion at this position is referred to as an effluent standard. Finally, square 4 represents criteria that are associated with sources of pollution other than pollutants. These criteria might include measures such as flow timing and pattern (a hydrologic criterion), abundance of nonindigenous taxa, some quantification of channel modification (e.g., decrease in sinuosity), etc.

Because the designated use is stated in written and qualitative terms, the challenge is to logically relate the criterion to the designated use. Establishing this relationship is easier as the criterion moves closer to the designated use (Figure 2-1). In addition, the more precise the statement of the designated use, the more accurate the criterion will be as an indicator of that use. For example, the criterion of fecal coliform count may be used for determining if the use of water contact recreation is achieved, and the fecal count criterion may differ among waterbodies that have primary versus secondary water contact as their designated use.

Surrogate variables often are selected for use as criteria because they are easy to measure. Although the surrogate may have this appealing attribute, its usefulness can be limited if it cannot be logically related to a designated use. For example, chlorophyll *a* has been chosen as a biocriterion in some states because it is a surrogate for aesthetic conditions or the status of the larger aquatic ecosystem. In North Carolina, the ambient water quality standard of 40 µg/l for chlorophyll *a* was proposed for lakes, reservoirs, sounds, estuaries, and other slow-moving waters not designated as trout waters. However, a discussion of the appropriate designated uses for the waters of the state and how this criterion is logically related to those uses did not accompany the adoption of this criterion.

As with setting designated uses, the relationship among waterbodies and segments must be considered when determining criteria. For example, where a segment of a waterbody is designated as a mixing zone for a discharge, the criterion adopted should assure that the mixing zone use will not affect the attainment of the uses designated for the surrounding waterbody. In a similar vein, the desired condition of a small headwater stream may need to be chosen as it relates to other waterbodies in the watershed. Thus, an ambient nutrient criterion may be set in a small headwater stream to secure a designated use in a downstream estuary, even if there are no localized effects of the nutrients in the small headwater stream. Conversely, a higher fecal coliform criterion that supports only secondary contact recreation may be warranted for a waterbody with little likelihood of being a recreational resource—if the fecal load dissipates before the flow reaches an area designated for primary contact recreation.

DECISION UNCERTAINTY

Ambient-focused water quality management requires one to ask whether the designated use is being attained and, if not, the reasons for nonattainment and how the situation can be remedied. Neither of these questions, which make reference to the chosen criteria, can be answered with complete certainty. Determining use attainment requires making criterion measurements at different locations in the waterbody and at different times and comparing the measurements to the standard. Individual measurements of a single criterion constitute a sample, and statistical inference procedures use the sample data to test hypotheses about whether the actual condition in the water meets the criterion. Errors of inference are always possible in statistically valid hypothesis testing. It is possible to falsely conclude that a criterion is not being met when it is. It is also possible to conclude that a criterion is being met when in fact it is being violated. Chapter 3 includes recommendations for controlling and managing such uncertainty.

Water quality management also requires models to relate the criterion to activities that might control pollution. For example, a criterion requiring a certain DO level may be chosen to help meet the designated use of a trout fishery. Models will be required to relate a management practice, such as fertilizer control, to the DO criterion. These types of models can be broadly labeled as models that relate stressors (sources of pollutants and pollution) to responses—similar to models used in hazardous waste risk assessment and many other fields. Stressors include human activities likely to cause impairment, such as the presence of impervious surfaces in a watershed, cultivation of fields too close to the stream, over-irrigation of crops with resulting polluted return flows, the discharge of domestic and industrial effluent into waterbodies, dams and other channelization, introduction of nonindigenous taxa, and overharvesting of fishes. Indirect effects of humans include the clearing of natural vegetation in uplands that alters the rates of delivery of water and sediment to stream channels.

A careful review of direct and indirect effects of human activities suggests five major classes of environmental stressors: alterations in physical habitat, modifications in the seasonal flow of water, changes in the food base of the system, changes in interactions within the stream biota, and release of contaminants (conventional pollutants) (Karr, 1990; NRC, 1992). The presence of one of more of these in a landscape may be responsible for changes in a waterbody that result in failure to attain a designated use. Ideally, models designed to protect or restore water quality to ensure attainment of designated uses should include all five classes of pollution. The broad-based approach implicit in these five features is more likely to solve water resource problems because it requires a more integrative diagnosis of the cause of degradation (NRC, 1992).

Models that relate stressors to responses can be of varying levels of complexity (Chapter 4). Sometimes, models are simple conceptual depictions of the relationships among important variables and indicators of those variables, such as the statement "human activities in a watershed affect water quality including the condition of the river biota." More complicated models can be used to make predictions about the assimilative capacity of a waterbody, the movement of a pollutant from various point and nonpoint sources through a watershed, or the effectiveness of certain best management practices.

There are two significant sources of uncertainty in any water quality management program: epistemic and aleatory uncertainty (Stewart, 2000). Epistemic uncertainty—incomplete knowledge or lack of sufficient data to estimate probabilities—is a by-product of our reliance on models that relate sources of pollution to human health and biological responses. We are limited by incomplete conceptual understanding of the systems under study, by models that are necessarily simplified representations of the complexity of the natural and socioeconomic systems, as well as by limited data for testing hypotheses and/or simulating the systems. Limited conceptual understanding leads to parameter uncertainty. For example, at present there is scientific uncertainty about the parameters that can represent the fate and transfer of pollutants through watersheds and waterbodies. It is plausible to argue that more complete data and more work on model development can reduce epistemic uncertainty. Thus, a goal of water quality management should be to increase the availability of data, improve its reliability, and advance our modeling capabilities. Indeed, Chapter 4 describes ways in which improved data and modeling can narrow the band of uncertainty and ways to characterize the remaining uncertainty.

However, complete certainty in support of water quality management decisions cannot be achieved because of aleatory uncertainty—the inherent variability of natural processes. Aleatory uncertainty arises in systems characterized by randomness. For example, if a pair of dice is thrown, the outcome can be predicted to be between 2 and 12, although the exact outcome cannot be predicted. The example of the dice toss represents the best-case scenario of a system characterized by randomness, because it is a closed system in which we have complete confidence that the result will be between 2 and 12. Not only are waterbodies, watersheds, and their inhabitants characterized by randomness, but they are also open systems in which we cannot know in advance what the boundaries of possible biological outcomes will be.

Thus, uncertainty is a reality that water quality management must recognize and strive to assess and reduce when possible. It derives from the need to use models that relate actions taken to alter the stressors so that the desired criterion and designated use of a waterbody will be secured. Although the purpose of water quality modeling will change depending on how close to the designated use the criterion is positioned, the importance of modeling and the inevitable uncertainties of model results remain.

CONCLUSIONS AND RECOMMENDATIONS

The two major themes of this chapter represent areas in water quality management where science and public policy intersect. First, with respect to the setting of water quality standards, in order for designated uses to reflect the range of scientific information and social desires for water quality, there must be substantial stratification and refinement of designated uses. Information from science can and must be part of this process; however, there are unavoidable social and economic decisions to be made about the desired state for each waterbody. Second, although science should be one cornerstone of the program, an unwarranted search for scientific certainty is detrimental to the water quality management needs of the nation. Recognition of uncertainty and creative ways to make decisions under such uncertainty should be built into water quality management policy, as discussed in the remaining chapters.

1. Assigning tiered designated uses is an essential step in setting water quality standards. Clean Water Act goals (e.g., "fishable," "swimmable") are too broad to be operational as statements of designated use. However, designated uses will still remain narrative statements.

2. Once designated uses are defined, the criterion chosen to measure use attainment should be logically linked to the designated

use. The criterion can be positioned anywhere along the causal chain connecting stressors (sources of pollution) to biological response. As the designated uses are expressed with more detail and are appropriately tiered, the criterion can be more readily related to the use. However, criteria should not be adopted based solely on the ease of measurement in making this link.

3. Expectations for the contribution of "science" to water quality management need to be tempered by an understanding that uncertainty cannot be eliminated. In both the assessment and planning processes, even the best available tools cannot banish uncertainty stemming from the variability of natural systems.

REFERENCES

Karr, J. R. 1990. Bioassessment and Non-Point Source Pollution: An Overview. Pages 4-1 to 4-18 in Second National Symposium on Water Quality Assessment. Washington, DC: EPA Office of Water.

National Research Council (NRC). 1992. Restoration of Aquatic Ecosystems. Washington, DC: National Academy Press.

Stewart, T. R. 2000. Uncertainty, judgment, and error in prediction. In Prediction: Science, Decision Making, and the Future of Nature. D. Sarewitz, R. A. Pielke Jr., and R. Byerly Jr., eds. Washington, DC: Island Press.

3
Waterbody Assessment: Listing and Delisting

On July 27, 2000, the Assistant Administrator for Water at the U. S. Environmental Protection Agency (EPA) testified before a U.S. House committee that over 20,000 waterbodies across the United States were not meeting water quality standards according to Section 303d lists. Because of legal, time, and resource pressures placed upon the states and EPA, there is considerable uncertainty about whether many of the waters on the 1998 303d lists are truly impaired. In many instances, waters previously presented in a state's 305b report[5] or evaluated under the 319 Program[6] were carried over to the state's 303d list without *any* supporting water quality data [e.g., see Iowa Senate File 2371, Sections 7–12 (Credible Data Legislation)]. Meanwhile, some waters that may be impaired have yet to be identified and listed.

The creation of an accurate and workable list of impaired waters is dependent on the first three steps of the Total Maximum Daily Load (TMDL) process, as depicted in Figure 1-1. States need to decide what waters should be assessed in the first place, how to create water quality standards for those waters, and then how to determine exceedance of

---

[5] The Clean Water Act Section 305b report—the National Water Quality Inventory Report—is the primary vehicle for informing Congress and the public about general water quality conditions in the United States. This document characterizes water quality, identifies widespread water quality problems of national significance, and describes various programs implemented to restore and protect our waters (http://www.epa.gov/305b/).

[6] Under the Clean Water Act Section 319 Nonpoint Source Management Program, States, Territories, and Indian Tribes receive grant money to support a wide variety of activities, including technical assistance, financial assistance, education, training, technology transfer, demonstration projects, and monitoring to assess the success of specific nonpoint source implementation projects (http://www.epa.gov/owow/nps/cwact.html).

those standards. Ideally, all these activities are encompassed and coordinated under the umbrella of a holistic ambient water quality monitoring program, described in the next section. However, given resource constraints, the approaches currently used in most states to list impaired waters fall short of this ideal. In recognition of these constraints, the committee recommends changes to the TMDL program that would make the lists more accurate over the short and long terms. In addition, this chapter includes discussion on identifying waters to be assessed, defining measurable criteria for water quality standards, and interpreting monitoring results for making the listing (and delisting) decision.

ADEQUATE AMBIENT MONITORING AND ASSESSMENT

The demands of an ambient-focused water quality management program, such as the TMDL program, require changing current approaches toward monitoring and assessment and subsequent decision-making. In many states, administrative performance measures (e.g., number of TMDLs developed, number of permits issued, and timeliness of actions) have been the principal measure of program effectiveness (Box 3-1). Such administrative measures are important, but reliance on such measures diverts attention and resources away from environmental indicators of waterbody condition—the principal measures of effectiveness and success. Rather, information for decision-making should be based on carefully collected and interpreted monitoring data (Karr and Dudley, 1981; Yoder, 1997; Yoder and Rankin, 1998). The committee recognizes that state ambient monitoring programs have multiple objectives beyond the TMDL program (e.g., 305b reports, trends and loads assessments, and other legal requirements), which are not addressed in this report. It is suggested that to make efficient use of resources, states evaluate the extent to which their present ambient monitoring programs are coordinated and collectively satisfy their objectives.

Ambient monitoring and assessment begins with the assignment of appropriate designated uses for waterbodies and measurable water quality criteria that can be used to determine use attainment (EPA, 1995a). The criteria, which may include biological, chemical, and physical measures, define the types of data to be collected and assessed. In response to the Government Performance and Results Act, the EPA Office of Water has developed national indicators for surface waters (EPA, 1995a) and a conceptual framework for using environmental information in decision-making (EPA, 1995b). EPA's Office of Research and Development

**BOX 3-1
Ohio's Experience with TMDLs**

In 1998, Ohio EPA's Division of Surface Water (DSW) made recommendations for a process to develop TMDLs (Ohio EPA, 1999). The impetus for developing a comprehensive TMDL strategy was (1) the national attention brought about by lawsuits filed by environmental organizations and (2) the potential for the TMDL process to address all relevant sources of pollution to a waterbody. Prior to realizing the importance of this issue, state water quality management efforts were focusing on point sources and National Pollution Discharge Elimination System (NPDES) permitting, although since 1996, the leading cause of waterbody impairment has been shown to be nonpoint pollution and habitat degradation (Ohio EPA, 2000; Section 305b report).

An agreement was reached between Ohio EPA and U.S. EPA Region V on a 15-year schedule for TMDL development. Ohio's 1998 303d list shows 881 of 5,000 waterbody segments as being impaired or threatened in 276 of the 326 watershed areas. Thus, completing TMDLs for all the currently listed segments by 2013 (in keeping with the 15-year schedule) will require an average of 18 watershed TMDLs per year assuming that no new watersheds are added to future revisions of the 303d lists. It is understood that this latter assumption is unrealistic because a good portion of the state's 5,000 waterbody segments has yet to be assessed, and it is a near certainty that additional waterbodies and watersheds will be listed. Ohio recognizes that the technical and management processes required to implement TMDLs will need to go beyond the purview of the past emphasis on NPDES permits and point sources.

At present, Ohio estimates it has sufficient resources available to develop only half of the TMDLs needed each year to produce the quality of product needed to meet various program expectations and expectations of stakeholders. Using 1998 as a baseline, approximately 16 percent of the DSW's resources were dedicated to efforts that directly support TMDL development (see pie chart below). Without increases in funding, the resources will need to be diverted from other programs, or the pace of TMDL development will slow to the point where the 15-year schedule will need to be significantly extended. Diverting resources from other programs is highly unlikely in that each program faces unique challenges, including reduction and elimination of NPDES permit backlogs

continues

> **BOX 3-1 Continued**
>
> and the growing need for new source permits, both of which place new burdens on the largest share of DSW resources. Devoting additional resources to TMDL development and implementation would require significant changes in water quality management emphasis on the national level, which seems unlikely given historical inertia and the emphasis placed on permitting programs by EPA and the states. Better coordination between competing programs as well as additional resources are needed to resolve the present TMDL resource shortfall dilemma. Focusing water quality management more on environmental results (as opposed to administrative accomplishments alone) should provide a framework to better unify the emphasis and direction of competing programs.
>
>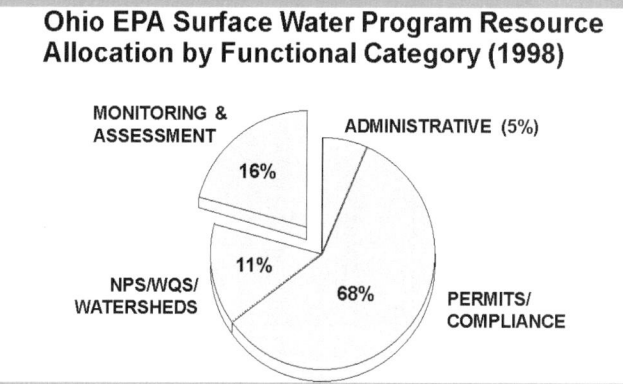

recently published technical guidelines for the evaluation of ecological indicators (Jackson et al., 2000). One set of measurable parameters, termed indicators in Table 3-1, is offered for illustration. The core indicators include baseline biological, chemical, and physical parameters that comprise the basic attributes of aquatic ecosystems supplemented by specific chemical, physical, and bacteriological parameters from water, sediment, and tissue media, depending on the applicable designated use(s) and watershed-specific issues. Additional indicators not listed (e.g., biochemical markers and whole toxicity testing) may be appropriate as the situation dictates.

TABLE 3-1. Core and Supplemental Indicators and Parameters that Comprise the Elements of an Adequate State Monitoring and Assessment Framework (after ITFM, 1992, and Yoder, 1997).

Core Indicators

Fish	Macroinvertebrates	Periphyton	Physical habitat	Chemical quality
Use at least two assemblages			• Channel morphology • Flow regime • Substrate quality • Riparian condition	• pH • Temperature • Conductivity • DO

For Specific Designated Uses, add the following:

	Aquatic Life	Recreation	Water Supply	Human/Wildlife Consumption
Base list	• Ionic strength • Nutrients, sediment	• Fecal bacteria • Ionic strength	• Fecal bacteria • Ionic strength • Nutrients, sediment	• Metals (in tissues) • Organics (in tissues)
Supplemental list	• Metals • Organics • Toxics	• Other pathogens • Organics	• Metals • Organics • Other pathogens	

More than one criterion may be necessary to determine attainment of a designated use, and each criterion will have strengths and limitations. In many instances of impairment—for example when riparian and aquatic habitats have been modified or flow regimes altered—biological parameters are better than chemical parameters at reflecting the condition of the aquatic ecosystem (Box 3-2). This is because biological assemblages respond to and integrate all relevant chemical, physical, and biological factors in the environment whether of natural or anthropogenic origin. On the other hand, relying only on biological assessments would not allow precise enough determination of associated causes and sources of impairments to satisfy water quality management needs including TMDL development. Over the long term, a full complement of measured parameters must be the goal for water quality monitoring, assessing

---

BOX 3-2
The Information Value of Monitoring Multiple Criteria

The tendency for misdiagnosis of impairment by relying on only one type of criterion was illustrated in a study of more than 2,500 paired stream and river sampling sites in Ohio (Ohio EPA, 1990; Rankin and Yoder, 1990). In 51.6 percent of the samples, the results from biomonitoring and chemical monitoring agreed—that is, they both detected either impairment or attainment of the water quality standard. This was particularly true for certain classes of chemicals (e.g., toxicants), where an exceedance as measured by the chemical parameter was always associated with a biocriteria impairment. However, in 41.1 percent of the samples, impairment was revealed by exceedance of the biocriteria but not by exceedance of the chemical criteria. These results suggest that impairment may go unreported in areas where only chemical measurements are made. Interestingly, in 6.7 percent of the samples, chemical assessment revealed impairment that was not detected by bioassessment (especially for parameters such as ammonia-N, dissolved oxygen (DO), and occasionally copper). This latter occurrence is likely related to the fact that biocriteria have been stratified to reflect regional or ecotype peculiarities, and the more generically derived chemical criteria have not. Both the under- and overprotective tendencies of a chemical-criteria-only approach to water quality management can be ameliorated by joint use of chemical criteria and biocriteria, each used within their most appropriate indicator roles and within an adequate monitoring and assessment framework.

chemistry and biology in a complementary manner and in their most appropriate indicator role (Karr, 1991; ITFM 1992, 1993, 1995; Yoder, 1997; Yoder and Rankin, 1998).

At present, monitoring resources available to some states often do not allow for collecting and interpreting data for such a comprehensive suite of parameters. Indeed, ITFM (1995) reported that of the funding allocated by state and federal agencies to water quality management activities, only 0.2 percent was devoted to ambient monitoring. GAO (2000) has also noted the lack of adequate state budgets for the collection of meaningful data and for data interpretation. In response to these resource shortfalls, the tendency has been to use only a single indicator of ambient conditions and often just a limited number of observations. Although some parameters can be monitored at lower costs than others, all monitoring can be costly (Yoder and Rankin, 1995).

After standards development, a second requirement is adoption of a strategic and consistent approach to sampling and assessment given limited data collection resources. Currently, the states use vastly different frameworks for monitoring and assessment, the net result of which is widely divergent estimates of the extent of impaired waters and of the proportion of waters that are fully assessed. This casts a great deal of uncertainty not only about what water quality problems are the most important, but also about the accuracy and completeness of their delineation. Errors in these estimates often become evident in the poor credibility of 303d listings.

A monitoring strategy that has promise in this limited-resource environment is the rotating basin approach, commonly referred to as a five-year basin approach (ITFM, 1995). As discussed in Box 3-3 for Florida, this approach is already followed by a number of states, at least in how ambient monitoring is accomplished[7]. As part of a rotating basin approach, individual waters are assessed at differing levels of complexity each year, allowing for localized problems to be identified and solutions to be developed. For example, whether an individual assessment consists of an initial screening to identify gross impairment or a full assessment with more serious consequences will depend on how the information is to be used (for 305b reports, 303d listing, or other water quality programs). Over time, different waterbodies are intensively studied as part of the rotation. Data collected can be used to support a number of differ-

---

[7] In some states, the rotating basin approach is considered to be part of the ambient monitoring program, while in others, it is a separate program. This report assumes the former throughout.

BOX 3-3
The Rotating Basin Program in Florida

Settlement of a lawsuit brought by Earthjustice against EPA for its failure to enforce timely actions to accomplish TMDL-related activities in Florida occurred in June 1999. Under the consent decree's (CD) "Terms of the Agreement," nearly 2,000 TMDLs in 711 waterbody segments are to be completed by the year 2011. Florida Department of Environmental Protection (FDEP) has been named the lead agency to produce and adopt TMDLs, but its efforts must be coordinated with numerous other state and local agencies. In addition, the state has created opportunities for public participation throughout the TMDL generation and adoption process.

To address the challenge of conducting the TMDL program and to better allocate its available resources, on July 1, 2000, Florida moved to the rotating basin approach for watershed management. Florida's rotating basin approach has five phases (see below), with each phase taking about one year to complete. Further, FDEP has divided the state into 30 areas based on 8-digit hydrologic unit codes (HUCs), such that six areas representing approximately one-fifth of Florida will be in the TMDL adoption phase in any one year. To meet the timelines ordered in the CD for Florida, FDEP must limit the time, effort, and resources it can commit in any one phase or waterbody.

Because EPA has largely focused on addressing point source discharges through the NPDES permitting program, state and local governments have in many cases taken the lead in dealing with nonpoint source issues, usually outside of the TMDL program. These programs often provide a flexible option to the time and budget constraints mentioned above. Florida believes that if local stakeholders are willing to initiate substantive programs that can fully, or even partially, accomplish the goals of the TMDL program at an expedited pace, then state and federal agencies should be able to support these actions, rather than delay or resist them. For example, in southwest Florida, a group of concerned stakeholders combined to form a "Nitrogen Consortium" (NC) to reduce inputs of nitrogen from all sources to the waters of Tampa Bay. Working together with the Tampa Bay Estuary Program and the FDEP, the NC developed a plan designed to "hold the line" against future increases of nitrogen (Tampa Bay National Estuary Program, 1996). Specific load-reduction efforts have been identified within the basin that allow for anticipated growth to occur without resulting in a net increase in nitrogen loads to Tampa Bay. As would be anticipated under the conditions of a more formal TMDL, periodic reviews are made of the underlying assumptions and models used to further refine the nitrogen loads and associated goals. Although FDEP has not formally adopted a TMDL for Tampa Bay, EPA has approved these "hold the line" limits as a TMDL for Tampa Bay.

continues

BOX 3-3 Continued

Florida's Basin Management Cycle: *5 phases*

	What happens in this phase?	**When does it occur?**
Phase I Preliminary Basin Assessment	Build basin management team Prepare Status Report - Document physical setting - Conduct water quality & TMDL assessments - Inventory existing & proposed management activities - Identify & prioritize management goals & objectives, & issues of concern - Develop Plan of Study	Years 1-2
Phase II Strategic Monitoring	Carry out strategic monitoring to collect additional data	Years 1-3
Phase III Data Analysis & TMDL Development	Compile & evaluate new data Finalize list of waters requiring TMDL Develop TMDL Identify additional data collection needs Report new findings	Years 2-4
Phase IV Management Action Plan	Finalize management goals & objectives Develop draft Management Action Plan Identify monitoring & management partnerships, needed rule changes, legislative actions, and funding opportunities Obtain participants' commitment to implement plan Develop Monitoring & Evaluation Plan	Years 4-5
Phase V Implementation	Implement Management Action Plan Secure project funding Carry out rule development/legislative action Transfer information to public & other agencies Conduct environmental education Monitor & evaluate implementation of plan	Year 5+

ent reporting and planning requirements, including a finding of attainment of water quality standards, a determination of impairment, or possible delisting if the waterbody is found not to be impaired. Initial assessments that identify a waterbody as *potentially* impaired could be followed up by more thorough assessment. The rotating basin approach is an iterative process where the end result is both continual improvement of water quality management tools and policies and the ability to respond to emerging issues.

Conclusions and Recommendations

1. To achieve the goal of ambient-based water quality management, monitoring and reporting must mature to focus on the condition of the environment as the principal measure of success rather than on administrative measures.

2. Biological parameters should be used in conjunction with physical and chemical parameters to assess the condition of waterbodies. The use of both biological and chemical parameters is needed because they provide different and complementary types of information about the source and extent of impairment.

3. Evidence suggests that limited budgets are preventing the states from monitoring for a full suite of indicators to assess the condition of their waters and from embracing a rotating basin approach to water quality management. Currently, EPA is assessing the sufficiency of state resources to develop and implement TMDLs. Depending on the results of that assessment, Congress might consider aiding the states, for example through matching grants to improve data collection and analysis. EPA would be instructed to develop guidelines for such a program, if needed, making eligibility contingent on an approved statewide monitoring and assessment strategy.

4. To allow states to better target limited monitoring budgets, EPA should set the TMDL calendar in concert with each state's rotating basin program. The rotating basin approach used by several states is an excellent example of a rigorous approach to ambient monitoring and data collection that can be used to conduct waterbody assessments of varying levels of complexity. For example, this approach can be used to create 305b reports, to list impaired waters, and to develop

TMDLs. Once TMDLs are developed, the rotating basin approach could allow state and local governments to issue permits and implement management programs based on the TMDLs in a coordinated manner.

DEFINING ALL WATERS

As shown in Figure 1-1, the TMDL process begins with identification of all waters for which achievement of water quality standards is to be assessed. The proposed regulations for the TMDL program (EPA, 1999a) define a waterbody as "a geographically defined portion of navigable waters, waters of the contiguous zone, and ocean waters under the jurisdiction of the United States, including segments of rivers, streams, lakes, wetlands, coastal waters and ocean waters." The proposed regulations also require that states identify the geographic location of listed waterbodies using a "nationally recognized georeferencing system as agreed to by [the state] and the EPA." States identify listed waterbodies using a variety of georeferencing systems, including stream segments in the EPA's reach file system and watersheds in the U. S. Geological Survey (USGS) system of hydrologic drainage basins. The use of such systems for documenting the location of listed waters is convenient and provides a degree of national standardization to the TMDL process. However, the selection of a georeferencing system and a spatial scale for defining the totality of state waters is a more complicated issue (aside from the policy issue of national standardization).

The EPA's definition of waterbody implies that all state waters should be considered in the search for impaired waters and provides no guidance on a practical upstream limit or spatial scale to observe in that search. In theory, the hierarchy of tributaries in a watershed extends upstream indefinitely. In practice, however, the choice of a lower limit on spatial scale or stream size has a very large influence on the total number of stream miles and small lakes that are included in the definition of state waters and thus require some form of assessment. For example, RF1, the original version of the EPA's national reach file system (DeWald et al., 1985) contained approximately 65,000 stream reaches totaling approximately 1 million km of stream channels. Now considered by EPA to be inadequate for describing the nation's river and stream system, RF1 has been replaced by the National Hydrography Dataset (NHD) containing more than 3 million reaches totaling nearly 10 million km of channels. Moreover, a number of states have petitioned the EPA to add still lower-order reaches (i.e., smaller streams) to the NHD in order to document the location of waters assessed by local interest groups. Because of local pressure and the lack of a regulatory lower limit on the size of streams and lakes to be considered, and because Geographic Information Systems (GIS) can document the existence and location of very small streams and lakes, the task of accurately and comprehensively assessing state waters has become formidable. At the current NHD scale, states contain an average of about 70,000 stream reaches (>100,000 km), and given recent trends, that average is rising.

This raises the question of how large the region of validity (the spatial area over which the data apply) is for data gathered at a single monitoring station. The question is conceptually troubling to begin with because the variability of water quality is large and continuous in both space and time. In practice, moreover, the de facto valid region for monitoring stations is extremely large. Given the spatially detailed treatment of rivers and streams in the NHD, however, most states would need to gather data from more than a thousand stations per year to maintain an average "monitoring ratio" of 100 km per station (assuming the NHD approximately describes state waters). This distance is clearly greater than the valid region for monitoring stations on most surface waters, especially because most of the channel length in state waters is contributed by relatively small streams (e.g., drainage areas less than 100 km^2) where water quality conditions may vary greatly over short distances. Thus, a substantial portion of state waters would appear to be located outside of the valid monitoring region for a state monitoring program of 1,000 stations. These waters are either left out of the decision process and are deemed not impaired by default, or they are included in the decision process with higher error rates.

One solution to this problem is to avoid the concept of a valid region for individual monitoring stations entirely and replace it with an approach in which monitoring data are used to develop statistical models of water quality in state waters. Water quality conditions at monitoring sites can be statistically related to known factors that cause impairment in watersheds (the size and location of stressors, for example), thus enabling estimates of water quality conditions at other unmonitored locations. As discussed later, this approach may also benefit the listing process.

Conclusions and Recommendations

1. Each state should develop a catalogue of waterbodies based on the National Hydrography Dataset for the purposes of defining state waters and designing sampling and assessment programs.

2. States should attempt to move away from the concept of a region of validity of individual monitoring stations and instead consider a statistical modeling approach to assessing the condition of waters. This approach would combine monitoring data with estimates of water quality based on statistical models.

DESIRABLE CRITERIA

This section considers the desired features of chemical and biological criteria as surrogates for designated use. For listing and delisting purposes, numeric and measurable criteria should be logically derived from the designated use statement. Ideally, appropriate designated uses and associated criteria are assigned to each waterbody prior to an assessment. Realistically, the cost and effort involved in categorizing every waterbody in advance of an assessment may be prohibitive, and many states' programs for setting appropriate use designation are continuing efforts. As is noted in Chapter 5, it is advisable to conduct a site-specific review to refine the standard once a waterbody is listed and before a TMDL is initiated.

One desired feature of a criterion is that it must be measurable with available monitoring methods. Unfortunately, federal guidelines for water quality assessment (EPA, 1994) do not assure this feature. In many cases there may be a discrepancy between the formulation of water quality criteria and the frequency with which water quality data are gathered.

A criterion may not be a single number, but instead may be represented as a frequency, duration, and magnitude. In the context of a pollutant, the *magnitude* refers to how much of the pollutant can be allowed in the water while still achieving the designated use. The magnitude can be chosen to protect against either acute or chronic effects of a pollutant. *Duration* refers to the period of time over which measurements of the pollutant are considered. Pollutant levels may be averaged over some number of hours or days to determine that amount of the pollutant that can be present without a loss of the designated use. The allowable *fre-*

quency at which the criterion can be violated (called an excursion) without a loss of the designated use also must be considered. Thus, in the case of a trout fishery, the criterion might specify a minimum DO (or maximum chlorophyll *a*) that can be realized for a period of time and the number of times this number can be violated before there is demonstrable harm to the designated use. It should be noted that these numbers are pollutant-specific, and they might vary with season depending on, for example, fish life-stage.

Establishing these three dimensions of the criterion is crucial for successfully developing water quality standards[8]. Currently, there are many cases where there are insufficient data collected in one or more of these three dimensions to evaluate attainment of water quality criteria. In addition, some standards are virtually impossible to comply with, especially when the frequency of allowable excursions is zero (called "no-exceedance" standards). Box 3-4 provides three examples of criteria that are either unmeasurable given current monitoring protocols or are exceedingly difficult to meet and thus constitute an intractable problem for the TMDL program. Careful consideration of the three dimensions of the criterion is also critical to the development of appropriate TMDLs. In the law, the letter "d" in TMDL refers to a *daily* load, which has been interpreted literally in some legal cases. However, for many pollutants, the load determined over a longer time period (e.g., a season or year) is more relevant to securing the designated use. Examples of this are nutrient and sediment criteria, where the duration component of the criterion is generally not stated as "daily."

A second desirable feature is that the measured criterion must be logically derived from the qualitative statement of the designated use. The closer the criterion is in the causal chain (Figure 2-1), the easier it is to make that connection. This has led to increased interest in biocriteria, particularly numeric measures of fish, benthic invertebrate, algal, and diatom assemblages. Recommendations to adopt biocriteria are often made because biocriteria integrate the effects of multiple stressors over time and space, thus minimizing the need for a large number of samples (Karr, 2000). A second advantage of using biocriteria is that, unlike chemical criteria, they are designed to be specific to certain regions and

---

[8] Specifying the magnitude, frequency, and duration is critical for chemical criteria, but may not be necessary for certain biological criteria. For example, the fecal coliform standard is best defined with all three components. On the other hand, many biocriteria such as IBI are well defined by a single number because they integrate biological, chemical, and physical effects over time.

> **BOX 3-4**
> **Problems Associated with Standards**
>
> Unmeasurable Standards
>
> By definition, the TMDL program requires that waterbodies meet water quality criteria daily, interpreted by some as meaning that the sampling frequency must be daily. This requires that a complete time series of grab or composite samples be taken daily without an interruption over a period of a minimum of three years. As one might expect, such time series of water quality data are almost never available for waterbody assessment (with the exception of the continuous monitoring for a few parameters such as DO or temperature). Samples are generally taken monthly for common parameters and annually or less often for some toxic chemicals that require expensive laboratory analytical methodology. Sediment sampling is done infrequently, perhaps once in a period of several years.
>
> Similarly, the frequency/duration components of water quality criteria for contact recreation are generally infeasible to measure. Many states use fecal coliform count as an indicator for the contact recreation. The standards are usually compared to the geometric mean of at least five samples taken over 30 days. This standard is not defined in terms of allowable excursions; thus, there is no frequency component. With the exception of waterbodies used for water supply, monitoring data are rarely collected often enough to comply with such a standard.
>
> *No-Exceedance Standard*
>
> Many states require that a numeric standard be maintained at all times, which implies that all monitored values of a parameter should be below the criterion. Such a limitation is a statistical impossibility because there is always a chance—albeit remote—that a water parameter may reach a high but statistically possible value exceeding an established standard. In addition, this requirement would seem to provide an incentive to sample as little as possible in order to reduce the chance of col-

conditions. For example, a swamp forest will typically violate DO criteria, and waterbodies in mountain areas with heavy metal-bearing rocks may violate heavy metal criteria. Biocriteria that are regionally relevant would not show those conditions as violations.

Fecal coliform counts and algal community parameters such as chlorophyll *a* are a type of biocriteria, but they are not comprehensive measures of waterbody condition. To make bioassessment more comprehen-

lecting a sample that is in exceedance. For example, it is possible that if nine samples are taken over a period of three years, none of the samples would, by chance, result in an excursion. If 100 samples are taken in the same period, a few (e.g., five or less) may exceed the standard. The former sampling scheme would indicate that the waterbody is in compliance while the other would not. Stream concentrations represent statistical time series for which only infinitesimally large values of a standard would have a 100 percent statistical probability of not ever being exceeded.

Flow Restriction Standards

To make "no-exceedance" standards easier to comply with, EPA (1992) and many states incorporated a flow restriction into the standards. Thus, the standards must be main-tained at all times except at flows that are less than some specified low flow value (one example is given below). Unfortunately, except for the "harmonic mean flow" (Singh and Ramamurthy, 1991), none of the critical low flows specified by EPA allow consideration of wet weather discharges (Novotny, 1999). Thus, under wet weather flows, the "no-exceedance" criterion is in effect. This ignores the fact that measured water quality parameters are naturally variable.

One type of flow restriction standard is based on hydrologically based design flows. To protect against acute effects, such water quality criteria must be met at all times except during the lowest daily flow occurring once every 10 years (referred to as 1Q10). To protect against chronic effects, water quality criteria must be met at all times except during the lowest flow occurring once every 10 years averaged over a 7-consecutive-day period (7Q10). This approach assumes that concentrations of pollutants of concern are decreasing as flows increase—likely to be true for the case of a continuous year-round discharge from a point source, but not for nonpoint sources. It should be noted that these design flows have "interim" status and were not recommended for general application with water quality standards. In addition, hydrologically based design flows vary from state to state.

sive, index systems have been developed that focus on characteristics of the biota expected in the particular region where the waterbody is located, including desired fish species and other associated organisms (Box 3-5).

The scientific community measures integrity by describing the biological condition of waterbodies that, as much as possible, have not been altered by human activity. When "pristine" or "minimally disturbed"

BOX 3-5
Index Systems for Bioassessment

During the past two decades, biological assessment—evaluating human-caused biotic changes apart from those occurring naturally—has become a part of water managers' tool kits. Two major approaches to ambient biological monitoring are used—the river invertebrate prediction and classification system (RIVPACS) and the multimetric index of biological integrity (IBI). Although their conceptual and analytical details differ, both RIVPACS and IBI (1) focus on biological endpoints to define waterbody condition, (2) use a concept of a regionally relevant reference condition as a benchmark, (3) organize sites into classes with similar environmental characteristics, (4) assess change and degradation caused by human effects, (5) require standardized sampling, laboratory, and analytical methods, (6) score sites numerically to reflect site condition, (7) define "bands," or condition classes, representing waterbody condition, and (8) furnish needed information for diverse management decisions (Karr and Chu, 2000).

RIVPACS was developed in England (Wright et al., 1989, 1997) with clones available for use in Australia (Norris et al., 1995) and Maine (Davies and Tsomides, 1997). IBI was developed in the United States (Karr, 1981; Karr et al., 1986; Karr and Chu, 1999) with clones applied by state and federal agencies (Ohio EPA, 1988; Davis et al., 1996; Barbour et al., 1999) and abroad (Hughes and Oberdorff, 1999). Although applications of RIVPACS are historically limited to invertebrates in rivers, IBI applications have been developed for diverse taxonomic groups and waterbody types. For example, a multimetric index (RFAI, reservoir fish assessment index) has been developed as a component of Tennessee Valley Authority's (TVA) "vital signs" monitoring program to assess fishery management success in reservoirs (Jennings et al., 1995; McDonough and Hickman, 1999).

As a general example, consider a minimally disturbed Pacific Northwest stream supporting self-sustaining populations of salmon and associated assemblages of invertebrates. With urban development, salmon decline and cutthroat trout become relatively more abundant, and certain invertebrate taxa (e.g., stoneflies) are reduced or eliminated. Tiered beneficial uses could in this case differentiate between streams supporting salmon vs. cutthroat trout, using an index based on the invertebrate assemblage as the biocriterion. Recent work in these streams suggests that a benthic index of biological integrity (B-IBI) of about 35 is a minimum required to maintain a healthy salmon population (Karr, 1998). If the IBI drops below 20 because of continued development, even the cutthroat trout will eventually disappear.

sites are used to define integrity, any site that has been altered by human actions must, by definition, lack integrity because its biota have changed in response to the actions of humans. For obvious reasons, reservoirs, farm ponds, and other waterbodies "created" by human actions cannot be assessed using this standard.

However, it does *not* follow that a waterbody lacking integrity is impaired or that restoring biological integrity is either possible or desirable. A waterbody that is described as lacking "biological integrity" should not be assumed to be in a less-than-desirable state. Rather, when a bioassessment finds that a waterbody diverges from integrity, there must be a social decision about whether that divergence is acceptable. In short,

> "The biota of minimally disturbed sites—those with integrity—provides a benchmark, a standard by which others are measured. The protection of that standard, or something very close to it, is likely to be the goal—the end toward which effort is directed—in relatively few places (e.g., national parks). The modern reality is that we are not able to preserve all areas in this benchmark condition. For example, restoring salmon to every Pacific Northwest stream is not realistic, yet a restoration goal that includes viable populations of cutthroat trout may be reasonable even in many urban or suburban streams. (Karr, 2000)

Measures of biological condition (e.g., IBI) inform society of the status of a water resource. But society must decide the desired designated use and then determine what level on the index numeric scale is, with reasonable certainty, likely to protect that designated use.

Recently, the EPA Office of Water has convened a working group of states and other supporting institutions to better define the gradient of biological condition from pristine to highly degraded and link this with operational measures such as numeric biocriteria in a manner that will ensure consistency across state programs. This is referred to as tiered aquatic life uses and is expressed as a biocondition axis. Examples of this framework already exist in Maine, Ohio, and Vermont. The expectation is that as states develop a more detailed system of tiered designated uses, they will also develop measurable biocriteria logically tied to those uses.

Conclusions and Recommendations

1. All chemical criteria and some biological criteria should be defined in terms of magnitude, frequency, and duration. Each of these three components is pollutant-specific and may vary with season. The frequency component should be expressed in terms of a number of allowed excursions in a specified period (return period) and not in terms of the low flow or an absolute "never to be exceeded" limit. The requirement of "no exceedances" for many water quality criteria is not achievable given natural variability alone, much less with the variability associated with discharges from point and nonpoint sources.

2. Water quality standards must be measurable by reasonably obtainable monitoring data. In many states, there is a fundamental discrepancy between the criteria that have been chosen to determine whether a waterbody is achieving its designated use and the frequency with which water quality data are collected.

3. Biological criteria should be used in conjunction with physical and chemical criteria to determine whether a waterbody is meeting its designated use. Biocriteria are more closely related to designated uses, they can be defined and measured, and they integrate the effects of multiple stressors over time and space.

LISTING AND DELISTING IN A DATA-LIMITED ENVIRONMENT

As discussed at the beginning of this chapter, states are confronted with lengthy lists of impaired waters requiring TMDLs, many of which were judged against inadequate standards or were not fully assessed as part of a comprehensive ambient monitoring program. This section proposes a mechanism for managing the large number of waters requiring attention by dividing the listing process into multiple smaller steps, as shown in Figure 3-1.

Figure 3-1 illustrates a framework for water quality management that is more detailed than the conceptualized steps of the TMDL process shown in Figure 1-1. Figure 3-1 begins with the identification of all waters to be assessed and the determination of appropriate water quality standards as in the current TMDL program. Following this, however,

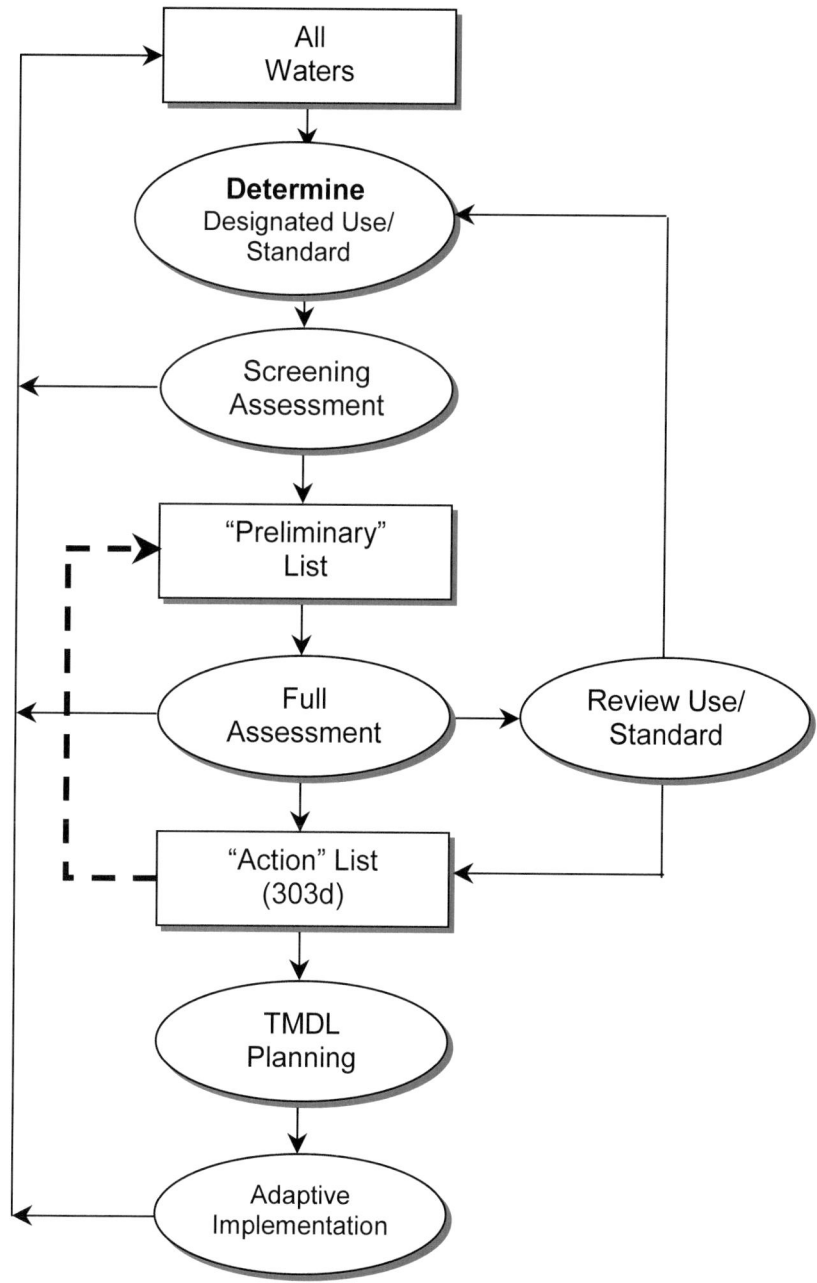

FIGURE 3-1 Framework for water quality management.

waters to be assessed would next go through an initial screening assessment. This involves comparing available, and often limited, data on water quality conditions with the existing applicable water quality criterion. If based on this initial screening assessment the waterbody is considered a candidate for impairment, it is advanced to the "preliminary" list for further consideration. It should be relatively easy to get on the preliminary list, the consequences of which include additional and immediate investigation to determine the nature and reality of a suspected problem. The term "preliminary" indicates that waterbodies on this list may later be placed on an action list, but they may also be declared unimpaired. Such a preliminary list has been suggested or employed in some states (e.g., Florida).

Those waterbodies placed on the preliminary list are the object of a more complete assessment that would involve additional monitoring and appropriate analysis of new data to reduce the uncertainty about their condition. If the decision from the full assessment is that the waterbody is impaired, then it moves to an "action list." One might think of the action list as the state's impaired waters (303d) list. The word "impaired" is a term of art. Impaired waters under Section 303d are analogous to "water quality limited segment(s)," as defined in the federal regulations (40 CFR Section 130.2(j)). The consequence of advancing to the action list is that additional resources are needed to either review and update the existing standard or complete a TMDL. (For those cases in which the existing criteria are not appropriate to a waterbody, Figure 3-1 allows for review of the water quality standard for that waterbody. The process for completing that review—use attainability analysis—is discussed in Chapter 5.)

The organizing concept in this idealized process is continuous and concurrent progress toward improved monitoring and listing decisions. The process moves forward from a position of limited information to more information; from uncertainty to more certainty; and from inaction to progressively larger and possibly more costly actions. Were EPA to endorse the idealized process represented in Figure 3-1, the listing process would be improved. For example, at the current time, there are thousands of waters on state 303d lists that were not placed there using adequate data or information. Waters in this category should be moved back to the preliminary list, represented by the dashed return arrow in Figure 3-1, to allow a more complete evaluation to be made.

Creating the Preliminary List

Determining whether there should be some minimum threshold of data available when evaluating waterbodies for attainment of water quality standards is an issue of great concern to states. On the one hand, many call for using only the "best science" in making listing decisions, while others fear many impaired waters will not be identified in the wait for additional data. The existence of a preliminary list addresses these concerns by focusing attention on waters suspected to be impaired without imposing on stakeholders and the agencies the consequences of TMDL development, until additional information is developed and evaluated.

In many cases, biological and limited water quality surveys along with an inventory of existing sources of pollution may provide adequate information for a screening assessment of the waterbody. Evaluated data are also an important source of information for determining if a waterbody should be placed on the preliminary list. Evaluated data may take many forms (e.g., data older than a certain age, beach closures based on fixed rainfall thresholds, visual observations, and statistical inferences from small data sets) and have been described differently from state to state[9]. In contrast, monitored data are viewed as being more comprehensive, typically using data less than five years old, and may include a wide array of direct measurements of water quality, including physical, chemical, or biological measures. Use of evaluated data has been controversial in water quality assessments under the Clean Water Act. The controversy would be lessened if the use of evaluated data were limited to placing waters on the preliminary list.

The quality of the data used to list waterbodies as impaired is frequently a concern. Beyond the normal data entry, sampling, and laboratory errors, states must determine the reliability of the data coming from a wide range of sources (especially for evaluated data). Some states have responded to this uncertainty by strictly limiting the data used in making assessments to those collected by the state's lead environmental agency or some other select group of data providers (such as USGS). To over-

---

[9] Evaluated data and/or information provides an indirect appraisal of water quality through such sources as information on historical adjacent land uses, aquatic and riparian health and habitat, location of sources, results from predictive modeling using input variables, and some surveys of fish and wildlife. Monitored data refers to direct measurements of water quality, including sediment measurements, bioassessments, and some fish tissue analyses (EPA, 1998, 2000).

come this uncertainty, and thereby expand the universe of reliable data, some states have required that associated meta data[10] be provided and entered into a central data repository (such as STORET).

Narrative criteria might also play a significant role in determining whether a waterbody should be placed on the preliminary list. Many water quality standards are characterized only by narrative criteria that express the desired target but do not allow comparison to a numeric value. For example, a typical narrative criterion for nutrients (nitrogen and phosphorus) in inland waters is "concentrations should be limited to the extent necessary to prevent nuisance growths of algae, weeds, and slimes" (as in New York State). Currently, violations based on interpretation of a narrative criterion may be a basis for placing a waterbody on the 303d list, even though such an evaluation is done without a numeric value of the criterion. EPA and the states have worked together over the last ten years to develop translators that will convert narrative standards to numeric criteria or guidance values (EPA, 1999b,c; NRC, 2000). While further progress is made in developing such translators, violations of narrative standards should be used to place waterbodies on the preliminary list.

The approaches to creating a preliminary list will vary from state to state. For example, in Florida, data and information used to place waters on the preliminary list have to meet certain basic QA/QC requirements as well as limited data sufficiency tests. Minimum sample sizes and confidence levels have been established, and both chemical and biological data are considered. States will have to decide upon and develop criteria for defining data sufficiency and analytical procedures for placing waterbodies on the preliminary list and the action list. EPA might be expected to assist in this process.

Moving Off the Preliminary List

Waters on the preliminary list should receive special monitoring attention. Movement from the preliminary list will be either back to the list of all waters *or* onto the action (303d) list. Movement off the preliminary list will demand a more analytically structured evaluation than

---

[10] Meta data is information about data and its usage, such as (1) what it is about, (2) where it is to be found, (3) how much it costs, (4) who can access it, (5) in what format it is available, (6) what the quality of the data is for a specified purpose, and (7) what spatial location and time period it covers.

was required for getting on the list. Each state should develop statistical procedures appropriate for testing attainment of each criterion. Sampling design, sample size, and QA/QC assurances for monitoring data would be defined, as would the appropriate tools for data analysis. If the data evaluated by the appropriate procedure indicate that there is no impairment, then delisting would follow. Delisting depends on analyses of sampling data and not on the implementation of a TMDL plan, although such a plan may be required to meet the criterion.

The process represented in Figure 3-1 is designed to improve the accuracy of the listing process. Placement of a waterbody on the preliminary list can serve as an indication to stakeholders that action should be taken soon to achieve water quality standards in order to avoid the costs associated with TMDL development. Because of the consequences of movement to the action list, there may be an incentive to keep waters on the preliminary list indefinitely. This incentive can be eliminated by requiring that a waterbody be automatically placed on the action (303d) list at the end of the next rotating basin cycle if additional analyses have not been undertaken. Such a requirement also may provide an incentive for point and nonpoint pollutant sources to contribute to the monitoring program in order to (potentially) avoid the consequences of a 303d listing.

Conclusions and Recommendations

1. EPA should approve the use of both a preliminary list and an action list instead of one 303d list. The two-list process would reduce the uncertainty that often accompanies a listing decision and would provide flexibility to the TMDL program.

2. If some waters on the current 303d list would be more appropriately catalogued on the preliminary list, EPA should allow states to move those waterbodies from the current 303d list to the preliminary list. If no legal mechanism exists to bring this about, Congress should create one. Many waters now on state 303d lists were placed there without the benefit of adequate data or waterbody assessment. These potentially erroneous listings contribute to a very large backlog of TMDL segments and foster the perception of a problem that is larger than it may actually be.

3. States should be allowed the flexibility to delist a waterbody without having to complete a TMDL if additional data or new in-

formation providing evidence of attainment of the water quality standard becomes available.

4. No waterbody should remain on the preliminary list for more than one rotating basin cycle. If the waterbody has not been removed from the preliminary list at the end of a rotating basin cycle, it should automatically be placed on the 303d list, unless EPA approves an exemption from such a requirement on a waterbody-by-waterbody basis. Criteria for granting exemptions could be developed by EPA.

5. To increase the reliability of the data used in listing waterbodies, EPA should require some limited amount of meta data for data submitted to STORET.

DATA EVALUATION FOR THE LISTING AND DELISTING PROCESS

Given finite monitoring resources, it is obvious that the number of sampling stations included in the state program will ultimately limit the number of water quality measurements that can be made at each station. Thus, in addition to the problem of defining state waters and designing the monitoring network to assess those waters, fundamental statistical issues arise concerning how to interpret limited data from individual sampling stations. Statistical inference procedures must be used on the sample data to test hypotheses about whether the actual condition in the waterbody meets the criterion. Thus, water quality assessment is a hypothesis-testing procedure.

A statistical analysis of sample data for determining whether a waterbody is meeting a criterion requires the definition of a null hypothesis; for listing a waterbody, the null hypothesis would be that the *water is not impaired*[11]. The analysis is prone to the possibility of both Type I error (a false conclusion that an unimpaired water is impaired) and Type II error (a false conclusion that an impaired water is not impaired). Different statistical analyses are needed depending on whether chemical or biological criteria are being assessed.

---

[11] For delisting, the null hypothesis might be that the water is impaired.

Statistical Approaches for Chemical Parameters

If chemical criteria—carefully designed to account for magnitude, frequency, and duration—are expected to be met, instantaneous measurements would be needed to determine compliance. Under current practice, however, even when states conduct frequent monitoring, sample sizes are limited, and so the possibility for false positive errors (Type I) and false negative errors (Type II) remains. As sample sizes increase, error rates can be better managed. For placement on the preliminary list, a small sample size may be acceptable. However, placement on the action list would require an increase in the number of sample points used in order to reduce the uncertainty in the listing and delisting decisions.

The committee does not recommend any particular statistical method for analyzing monitoring data and for listing waters. However, one possibility is that the *binomial hypothesis test* could be required as a minimum and practical first step (Smith et al., 2001). The binomial method is not a significant departure from the current approach—called the *raw score approach*—in which the listing process treats all sample observations as binary values that either exceed the criterion or do not, and the binomial method has some important advantages. For example, one limitation of the raw score approach is that it does not account for the total number of measurements made. Clearly, 1 out of 6 measurements above the criterion is a weaker case for impairment than is 6 out of 36. The binomial hypothesis test allows one to take sample size into account. By using a statistical procedure, sample sizes can be selected and one can *explicitly* control and make trade-offs between error rates (see Smith et al., 2001, and Gibbons, in press, for guidance on managing the risk of false positive and false negative errors)[12]. Several states, including Florida and Virginia, are considering or are already using the binomial hypothesis test to list impaired waters. Detailed examples of how to ap-

---

[12] The choice of a Type I error rate is based on the assessors willingness to falsely categorize a waterbody. It also is the case that, for any sample size, the Type II error rate decreases as the acceptable Type I error rate increases. The willingness to make either kind of mistake will depend on the consequences of the resulting actions (more monitoring, costs to do a TMDL plan, costs to implement controls, possible health risk) and who bears the cost (public budget, private parties, etc.). The magnitude and burden of a Type I versus Type II error depend on the statement of the null hypothesis and on the sample size. When choosing a Type I error rate, the assessor may want to explicitly consider these determinants of error rates.

ply this test are beyond the scope of this document, but can be found in Smith et al. (2001) and the proposed Chapter 62-303 of the Florida Administrative Code[13].

Whether the binomial or the raw score approach is used, there must be a decision on an acceptable frequency of violation for the numeric criterion, which can range from 0 percent of the time to some positive number. Under the current EPA approach, 10 percent of the sample measurements of a given pollutant made at a station may exceed the applicable criterion without having to list the surrounding waterbody. The choice of 10 percent is meant to allow for uncertainty in the decision process. Unfortunately, simply setting an upper bound on the percentage of measurements at a station that may violate a standard provides insufficient information to properly deal with the uncertainty concerning impairment.

The choice of acceptable frequency of violation is also supposed to be related to whether the designated use will be compromised, which is clearly dependent on the pollutant and on waterbody characteristics such as flow rate. A determination of 10 percent cannot be expected to apply to all water quality situations. In fact, it is inconsistent with federal water quality criteria for toxics that specify allowable violation frequencies of either one day in three years, four consecutive days in three years, or 30 consecutive days in three years (which are all less than 10 percent). Embedded in the EPA raw score approach is an implication that 10 percent is an acceptable violation rate, which it may not be in certain circumstances.

Both the raw score and binomial approaches require the analyst to "throw away" some of the information found in collected data. For example, if the criterion is 1.0, measurements of 1.1 and 10 are given equal importance, and both are treated simply as exceeding the standard. Thus, a potentially large amount of information about the likelihood of impairment is simply discarded. (The standard deviation can be used to set priorities for TMDL development or other restoration activities.) There are other approaches that are more effective at extracting information from a single monitoring sample, thereby reducing the number of samples needed to make a decision with the same level of statistical confi-

---

[13] This proposed rule chapter was approved for adoption by the Florida Department of Environmental Protection's Environmental Regulation Commission on April 26, 2001, but has not been officially filed for adoption by the Department because of a pending rule challenge before the Division of Administrative Hearings.

dence. For example, Gibbons (in press) suggests testing the data for normality or log normality and then examining the confidence intervals surrounding the estimated 90th percentile of the chosen distribution. When the data are neither normal nor lognormal, or when more than 50 percent of the observations are censored (below the detection limit), Gibbons suggests constructing a nonparametric confidence limit based on the binomial distribution of ranked data. Another approach that uses all the data to make a decision is "acceptance sampling by variables" (Duncan, 1974). In general, alternative statistical approaches transform questions about the proportion of samples that exceed a standard into questions about the center (or another parameter) of a continuous distribution. It should be noted that new approaches will bring new analytical requirements that must be taken into consideration. For example, if there is a requirement to specify a distribution, sufficient data must be available. In some cases, data from other similar sites may be needed to give an overall assessment of distribution type. Finally, as more powerful statistical procedures are used, water quality assessors will need to understand how to run the tests and also how to state hypotheses that clearly relate to the water quality criterion.

Statistical Approaches for Biological Parameters

Error bands exist with any sampled data, including bioassessment results. Thus, bioassessment procedures must also be designed to be statistically sound. The utility of any measure of stream condition depends on how accurately the original sample represents the condition in the stream—that is, how successful it is in avoiding statistical "bias." Protocols to for making such measurements are established in the technical literature (Karr and Chu, 1999) as well as in guidance manuals produced by EPA (Barbour et al., 1996, 1999; EPA, 1998a; Gibson et al., 2000).

There are three principal ways variability is dealt with in the process of deriving and using biocriteria (Yoder and Rankin, 1995). First, variability is compressed through the use of multimetric evaluation mechanisms such as IBI. Reference data for each metric are compressed into discrete scoring ranges (i.e., 5, 3, and 1). Second, variability is stratified via tiered uses, ecoregions, stream size categories (headwaters, wadable, boatable), and method of calibrating each metric (i.e., vectoring expectations by stream size). Third, variability is controlled through standardized operating procedures, data quality objectives (i.e., level of taxonomy), index sampling periods (to control for seasonal effects), replication of sampling, and training (Yoder and Rankin, 1995). One can, for example, avoid seasonal variation by carefully defining index sampling periods or variation among microhabitats by sampling the most representative microhabitat (Karr and Chu, 1999). Box 3-6 presents results of several studies in which the error around biological parameters was assessed.

BOX 3-6
Understanding Sources of Variability in Bioassessment

Sources of error evaluated in one study of biological monitoring data from New England lakes (Karr and Chu, 1999) included three types of variance: interlake variability (differences among lakes); intralake variability (variability associated with sampling different sites within a lake as decided by the field crew), and lab error (error related to subsample work in the lab). The interlake variability was the effect of interest, and the goal was to determine if that source of variability was dominant. Distribution of variance varied as a function of biological metric selected. Those measures with reduced variance except for the context of interest (e.g., interlake variability) were selected for inclusion in IBI to increase the probability of detecting and understanding the pattern of interest.

Two other studies involved an examination not of the individual metrics, but of the overall IBI (i.e., after individual metrics were tested and integrated into an IBI). For Puget Sound streams, 9 percent of variation came from differences within streams and 91 percent was variability across streams (reported in Karr and Chu, 1999, Fig. 35). For a study in Grand Teton National Park, streams were grouped in classes reflecting different amounts of human activity in their watersheds. In this case, 89 percent of the variance came from differences among the groups, and 11 percent came from differences among members of the same group (reported in Karr and Chu, 1999).

In all these cases, the goal was to find ways of measuring that emphasize differences among watersheds with differing human influences, while keeping other sources of variation small. Success in these examples was based on the development of an earlier understanding of sources of variation and then establishing sampling protocols that avoid other irrelevant sources of variation (such as variation stemming from the differing abilities of personnel to select and use methods). If these sources of variation are controlled for, then the study can emphasize the kind of variation that is of primary interest (e.g., human influence gradients).

Waterbody Assessment: Listing and Delisting 61

Conclusions and Recommendations

1. EPA should endorse statistical approaches to proper monitoring design, data analysis, and impairment assessment. For chemical parameters, these might include the binomial hypothesis test or other statistical approaches that can more effectively make use of the data collected to determine water quality impairment than does the raw score approach. For biological parameters, these might focus on improvement of sampling designs, more careful identification of the components of biology used as indicators, and analytical procedures that explore biological data as well as integrate biological information with other relevant data.

2. States should be required to report the statistical properties of the sample data analyses used to make listing determinations. Error rates, confidence limits, or other means of conveying uncertainty should be presented along with the rationale for a decision to list or delist a waterbody.

USE OF MODELS IN THE LISTING PROCESS

As stated in EPA guidance documents as well as the Federal Advisory Committee Act (FACA) report (EPA, 1998b), monitoring data are the preferred form of information for identifying impaired waters. Model predictions might be used in addition to or instead of monitoring data for two reasons: (1) modeling could be feasible in some situations where monitoring is not, and (2) integrated monitoring and modeling systems could provide better information than monitoring alone for the same total cost. EPA guidance and the FACA report explicitly recognize the obvious practicality of the first reason, but largely ignore the potential importance of the second. This section considers some of the ways in which modeling might be used as a complement to monitoring and points out some limitations of modeling in informing the listing process.

Often, in attempting to estimate the frequency of violation of a standard, the number of pollutant concentration measurements made in a waterbody is so small that it is difficult to avoid false negative error with the desired level of confidence. One way in which a simple statistical model may assist in interpreting monitoring data in such cases is by introducing a variable to the analysis that is correlated with pollutant concentration. One common correlate of many water quality time series is stream flow, which is measured continuously at many monitoring stations, including nearly all USGS stations. The statistical methods for taking advantage of correlated stream flow data are called record extension techniques, several of which have been described and compared by Hirsch (1982). By modeling pollutant concentration as a function of streamflow and using the resulting model to estimate a denser concentration time series, a better estimate of the frequency distribution of pollutant concentration may be obtained. The predicted concentration time series then may be tested for violation frequency using either the binomial approach (see above) or the quantile approach. The value of this modeling approach over using pollutant data alone is directly dependent on the level of correlation that exists between the pollutant concentration and stream flow. Further discussion of the specific extension technique called MOVE (Maintenance of Variance–Extension) appears in Helsel and Hirsch (1991).

The EPA guidance on 303d listing suggests that a simple, but useful, modeling approach that may be used in the absence of monitoring data is "dilution calculations," in which the rate of pollutant loading from point sources in a waterbody (recorded as kg per day in NPDES permits, for example) is divided by the stream flow distribution to give a set of estimated pollutant concentrations that may be compared to the state standard. Simple dilution calculations assume conservative movement of pollutants through a watershed and ignore the fact that for most pollutants some loss of mass occurs during transport due to a variety of processes including evaporation, settling, or biochemical transformation (see, for example, Novotny and Olem, 1994). Thus, the use of dilution calculations will tend to bias the decision process toward false positive conclusions. Lacking a clear rationale for such a bias, a better approach would be to include a best estimate of the effects of loss processes in the dilution model.

Section 303d and related guidance from EPA emphasize the importance of searching for information on waterbodies that are suspected of violating water quality standards, which is understandable given the desire to limit the number of sites sampled and hence the cost of monitoring. Targeted monitoring will often increase the efficiency of the assessment process (i.e., reduce the total number of decision errors), but may have somewhat hidden effects on the balance of false positive and false negative errors. Targeted monitoring represents the informal use of a prior *probability distribution on impairment* to guide monitoring toward sites located in a particular region of the distribution. One of the most potentially valuable uses of modeling in relation to 303d listing would be to formalize the use of prior information on impairment probability in order to better organize the decision process. That is, modeling techniques such as SPARROW (Smith et al., 1997) could be used to estimate preliminary impairment distributions for all waterbodies in the state. These distributions would then be used to guide monitoring and control the rates of false positive and false negative error either through Bayesian or other methods of interpreting monitoring data. Limited monitoring resources generally could be focused on the sites where impairment was most *un*certain (i.e., where the estimated probability of impairment was neither very high nor very low), potentially improving the efficiency of monitoring. Sites at the extremes of the impairment distributions (i.e., extremely likely or unlikely to be impaired) would be less frequently monitored. Decisions for placing waters on a preliminary list might be made primarily on the basis of such modeling. (Formal placement of a waterbody on the 303d list would require additional monitoring.)

Conclusions and Recommendations

1. Models that can fill gaps in data have the potential to generate information that will increase the efficiency of monitoring and thus increase the accuracy of the preliminary listing process. For example, regression analyses that correlate pollutant concentration with some more easily measurable factor could be used to extend monitoring data for preliminary listing purposes. Models can also be used in a Bayesian framework to determine preliminary probability distributions of impairment that can help direct monitoring efforts and reduce the quantity of monitoring data needed for making listing decisions at a given level of reliability.

REFERENCES

Barbour, M. T., J. B. Stribling, J. Gerritsen, and J. R. Karr. 1996. Biological Criteria: Technical Guidance for Streams and Small Rivers. Revised Edition. EPA 822-B-96-001. Washington, DC: EPA Office of Water.

Barbour, M. T., J. Gerritsen, B. D. Snyder, and J. B. Stribling. 1999. Rapid Bioassessment Protocols for Use in Streams and Wadeable Rivers: Periphyton, Benthic Macroinvertebrates and Fish, Second Edition. EPA 841-B-99-002. Washington, DC: EPA Office of Water.

Davis, W. S., B. D. Snyder, J. B. Stribling, and C. Stoughton. 1996. Summary of State Biological Assessment Programs for Streams and Rivers. EPA 230-R-96-007. Washington, DC: EPA Office of Policy, Planning, and Evaluation.

DeWald, T., R. Horn, R. Greenspun, P. Taylor, L. Manning, and A. Montalbano. 1985. STORET Reach Retrieval Documentation. Washington, DC: EPA.

Environmental Protection Agency (EPA). 1994. Water Quality Standards Handbook. Second Edition. EPA 823-B-94-005a. Washington, DC: EPA Office of Water.

EPA. 1995a. Environmental indicators of water quality in the United States. EPA 841-R-96-002. Washington, DC: Office of Policy, Planning, and Evaluation.

EPA. 1995b. A conceptual framework to support development and use of environmental information in decision-making. EPA 239-R-95-012. Washington, DC: Office of Policy, Planning, and Evaluation.

EPA. 1998a. Lake and Reservoir Bioassessment and Biocriteria: Technical Guidance Document. EPA 841-B-98-007. Washington, DC: EPA Office of Water.

EPA. 1998b. Report of the FACA Committee on the TMDL Program. EPA 100-R-98-006. Washington, DC: EPA Office of the Administrator.

EPA. 1999a. Draft Guidance for Water Quality-based Decisions: The TMDL Process (Second Edition). Washington, DC: EPA Office of Water.

EPA. 1999b. Protocol for Developing Sediment TMDLs. First Edition. EPA 841-B-99-004. Washington, DC: EPA Office of Water.

EPA. 1999c. Protocol for Developing Nutrient TMDLs. First Edition. EPA 841-B-99-007. Washington, DC: EPA Office of Water.

General Accounting Office (GAO). 2000. Water Quality-Key EPA and State Decisions Limited by Inconsistent and Incomplete Data. GAO/RCED-00-54. Washington, DC: GAO.

Gibbons, R. D. In Press. An alternative statistical approach for performing water quality impairment assessments under the TMDL program.

Gibson, G. R., M. L. Bowman, J. Gerritsen, and B. D. Snyder. 2000. Estuarine and Coastal Marine Waters: Bioassessment and Biocriteria Technical Guidance. EPA 822-B-00-024. Washington, DC: EPA Office of Water.

Helsel, D. R., and R. M. Hirsch. 1992. Statistical Methods in Water Resources. Amsterdam: Elsevier. 522 p.

Hirsch, R. M. 1982. A comparison of four record extension techniques. Water Resources Research 15:1781–1790.

Hughes, R. M., and T. Oberdorff. 1999. Applications of IBI concepts and metrics to waters outside the United States and Canada. Pages 79–93 in T. P. Simon, editor. Assessing the Sustainability and Biological Integrity of Water Resources Using Fish Communities. Boca Raton, FL: CRC Press.

ITFM (Intergovernmental Task Force on Monitoring Water Quality). 1992. Ambient water quality monitoring in the United States: first year review, evaluation, and recommendations. Washington, DC: Interagency Advisory Committee on Water Data.

ITFM. 1993. Ambient water quality monitoring in the United States: second year review, evaluation, and recommendations. Washington, DC: Interagency Advisory Committee on Water Data.

ITFM. 1995. The strategy for improving water-quality monitoring in the United States. Final report of the Intergovernmental Task Force on Monitoring Water Quality. Washington, DC: Interagency Advisory Committee on Water Data.

Jackson, L. E., J. C. Kurtz, and W. S. Fisher, editors. 2000. Evaluation Guidelines for Ecological Indicators. EPA/620/R-99/005. Research Triangle Park, NC: EPA Office of Research and Development.

Jennings, M. J., L. S. Fore, and J. R. Karr. 1995. Biological monitoring of fish assemblages in Tennessee Valley Reservoirs. Regulated Rivers: Research and Management 11:263–274.

Karr, J. R. 1981. Assessment of biotic integrity using fish communities. Fisheries 6(6):21–27.

Karr, J. R. 1991. Biological Integrity: A Long-Neglected Aspect of Water Resource Management. Ecological Applications 1:66–84.

Karr, J. R. 1998. Rivers as sentinels: Using the biology of rivers to guide landscape management. Pages 502–528 in R. J. Naiman and R. E. Bilby, eds. River Ecology and Management: Lessons from the Pacific Coastal Ecosystems. New York: Springer.

Karr, J. R. 2000. Health, integrity, and biological assessment: The importance of whole things. Pages 209–226 in D. Pimentel, L. Westra, and R. F. Noss, editors. Ecological Integrity: Integrating Environment, Conservation, and Health. Washington, DC: Island Press. Pgs. 214–215.

Karr, J. R., and D. R. Dudley. 1981. Ecological perspective on water quality goals. Environmental Management 5:55–68.

Karr, J. R., and E. W. Chu. 1999. Restoring Life in Running Waters: Better Biological Monitoring. Washington, DC: Island Press.

Karr, J. R., and E. W. Chu. 2000. Sustaining living rivers. Hydrobiologia 422/423:1–14.

Karr, J. R., K. D. Fausch, P. L. Angermeier, P. R. Yant, and I. J. Schlosser. 1986. Assessment of biological integrity in running waters: A method and its rationale. Illinois Natural History Survey Special Publication 5. Illinois Natural History Survey, Urbana, Illinois.

McDonough, T. A., and G. D. Hickman. 1999. Reservoir fish assemblage index development: A tool for assessing ecological health in Tennessee Valley Authority impoundments. Pages 523–540 in T. P. Simon, editor. Assessing the Sustainability and Biological Integrity of Water Resources Using Fish Communities. Boca Raton, FL: CRC Press.

National Research Council (NRC). 2000. Watershed Management for Potable Water Supply: Assessing the New York City Strategy. Washington, DC: National Academy Press.

Novotny, V. 1999. Integrating diffuse/nonpoint pollution control and water body restoration into watershed management. Journal AWRA 35(4):717–727.

Novotny, V., and H. Olem. 1994. Water Quality: Prevention, Identification and Management of Diffuse Pollution. New York: Van Nostrand - Reinhold (distributed by Wiley).

Ohio EPA (Environmental Protection Agency). 1988. Biological Criteria for the Protection of Aquatic Life, volumes 1–3. Columbus, OH: Ohio EPA Ecological Assessment Section, Division of Water Quality Monitoring and Assessment.

Ohio EPA. 1990. Ohio water resource inventory, volume I, summary, status, and trends. Rankin, E. T., Yoder, C. O., and Mishne, D. A. (eds.). Columbus, OH: Ohio EPA Division of Water Quality Planning and Assessment

Ohio EPA. 1999. Total maximum daily load TMDL team report. Columbus, OH: Ohio EPA Division of Surface Water. 139 pp.

Ohio EPA. 2000. Ohio EPA five-year surface water monitoring strategy: 2000–2004 (draft). Columbus, OH: Ohio EPA Division of Surface Water, Ecological Assessment Unit.

Plafkin, J. L., M. T. Barbour, K. D. Porter, S. K. Gross and R. M. Hughes. 1989. Rapid Bioassessment Protocol for Use in Stream and Rivers: Benthic Macroinvertebrates and Fish. EPA 440/4-89/001. Washington, DC: EPA.

Rankin, E. T., and C. O. Yoder. 1990. A comparison of aquatic life use impairment detection and its causes between an integrated, biosurvey-based environmental assessment and its water column chemistry subcomponent. Appendix I, Ohio Water Resource Inventory (Volume 1). Columbus, OH: Ohio EPA, Division of Water Quality Planning Assessment. 29 pp.

Singh, K. P., and G. S. Ramamurthy. 1991. Harmonic Mean Flows for Illinois Streams. Champaign, IL: Illinois State Water Survey.

Smith, R. A., G. E. Schwarz, and R. B. Alexander. 1997. Regional interpretation of water-quality monitoring data. Water Resources Research 33(12):2781–2798.

Smith, E. P., K. Ye, C. Hughes, and L. Shabman. 2001. Statistical assessment of violations of water quality standards under Section 303(d) of the Clean Water Act. ES&T 35:606–612.

Tampa Bay National Estuary Program. 1996. Charting the Course—The Comprehensive Conservation and Management Plan for Tampa Bay.

Wright, J. F., D. Moss, R. T. Clarke, and M. T. Furse. 1997. Biological assessment of river quality using the new version of RIVPACS (RIVPACS III). Pages 102–108 in P. J. Boon and D. L. Howell (eds). Freshwater Quality: Defining the Indefinable? Scottish Natural Heritage, Edinburgh.Norris, R. H., B. T. Hart, M. Finlayson, and K. R. Norris (eds).

Wright, J. F., P. D. Armitage, and M. T. Furse. 1989. Prediction of invertebrate communities using stream measurements. Regulated Rivers: Research and Management 4:147–155.

Yoder, C. O. 1997. Important elements and concepts of an adequate state watershed monitoring and assessment program. ASIWPCA Standards & Monitoring, Ohio EPA Tech. Bull. MAS/1997-7-1. Columbus, OH: Ohio EPA Division of Surface Water.

Yoder, C. O., and E. T. Rankin. 1995. Biological criteria program development and implementation in Ohio, pp. 109–144, in W. Davis and T. Simon (eds.). Biological Assessment and Criteria: Tools for Water Resource Planning and Decision Making. Boca Raton, FL: Lewis Publishers.

Yoder, C. O., and E. T. Rankin. 1998. Biological response signatures and the area of degradation value: new tools for interpreting multimetric data, pp. 263–286. in W. Davis and T. Simon (eds.). Biological Assessment and Criteria: Tools for Water Resource Planning and Decision Making. Boca Raton, FL: Lewis Publishers.

4
Modeling to Support the TMDL Process

This chapter addresses the planning step (Figure 1-1) that occurs once a waterbody is formally listed as impaired. The main activity required during the planning step is an assessment of the relative contribution of different stressors (sources of pollution) to the impairment. For example, during this step Total Maximum Daily Loads (TMDLs) are calculated for the chemical pollutant (if there is one) causing the impairment, and the maximum pollutant loads consistent with achieving the water quality standard are estimated. Pollutant load limits alone may not secure the designated use, however, if other sources of pollution are present. Changes in the hydrologic regime (such as in the pattern and timing of flow) or changes in the biological community (such as in the control of alien taxa or riparian zone condition) may be needed to attain the designated use, as discussed in Chapter 2. As hydrologic, biological, chemical, or physical conditions change, the estimation of the TMDL can change.

Because they represent our scientific understanding of how stressors relate to appropriate designated uses, models play a central role in the TMDL program. Models are the means of making predictions—not only about the TMDL required to achieve water quality standards, but also about the effectiveness of different actions to limit pollutant sources and modify other stressors to reach attainment of a designated use. This chapter discusses the necessity for, and limitations of, models and other predictive approaches in the TMDL process. Thus, it directly addresses the committee's charge of evaluating the TMDL program's information needs and the methods used to obtain information.

MODEL SELECTION CRITERIA

Mathematical models can be characterized as empirical (also known as statistical) or mechanistic (process-oriented), but most useful models have elements of both types. An empirical model is based on a statistical fit to data as a way to statistically identify relationships between stressor and response variables. A mechanistic model is a mathematical characterization of the scientific understanding of the critical biogeochemical processes in the natural system; the only data input is in the selection of model parameters and initial and boundary conditions. Box 4-1 presents a simple explanation of the difference between the two types of models.

Water quality models for TMDL development are typically classified as either watershed (pollutant load) models or as waterbody (pollutant response) models. A watershed model is used to predict the pollutant load to a waterbody as a function of land use and pollutant discharge; a waterbody model is used to predict pollutant concentrations and other responses in the waterbody as a function of the pollutant load. Thus, the waterbody model is necessary for determining the TMDL that meets the water quality standard, and a watershed model is necessary for allocating the TMDL among sources. Some comprehensive modeling frameworks [e.g., BASINS (EPA, 2001) and Eutromod (Reckhow et al., 1992)] include both, but most water quality models are of one or the other type. Except where noted, the comments in this chapter reflect both watershed and waterbody models; examples presented may address one or the other model type as needed to illustrate concepts.

Although prediction typically is made with a mathematical model, there are certainly situations in which expert judgment can and should be employed. Furthermore, although in many cases a complex mathematical model can be developed, the model best suited for the situation may be relatively simple, as noted in examples described later in the chapter. Indeed, reliance on professional judgment and simpler modeling will be acceptable in many cases, and is compatible with the adaptive approach to TMDLs described in Chapter 5.

Highly detailed models are expensive to develop and apply and may be time consuming to execute. Much of the concern over costs of TMDLs appears to be based on the assumption that detailed modeling techniques will be required for most TMDLs. In the quest to efficiently allocate TMDL resources, states should recognize that simpler analyses can often support informed decision-making and that complex modeling studies should be pursued only if warranted by the complexity of the

> **BOX 4-1**
> **Mechanistic vs. Statistical Models**
>
> Suppose a teacher is conducting a lesson on measurements and sets out to measure and record the height and weight of ea`ch student. Unfortunately, the scale breaks after the first several children have been weighed. In order to proceed with the lesson (though on a somewhat different tack), a *mechanistically* inclined teacher might decide to use textbook data on the density of the human body, together with a variety of length measurements of each child (e.g., waist, leg, and arm dimensions), to estimate body volumes as the sum of the volumes of body parts. The teacher may then obtain the weights of the students as the product of density and volume. A *statistically* inclined teacher, on the other hand, might simply use the data obtained for the first several children in a regression model of weight on height that could then be used to predict the weights of the other students based on their height.
>
> The accuracy and utility of each of these two approaches depend on both the details of the input data and the calculation procedures. If the mechanistic teacher has good information on tissue densities, for example, and has the time to make many length measurements, the results may be quite good. Conversely, the statistical approach may yield quite acceptable results at a fraction of the mechanistic effort if enough children had been weighed before the scale broke, and if those children were approximately representative of the whole class in terms of body build. Moreover, the regression model comes with error statistics for its predictions and parameters. Although the same statistical approach would work with other groups of students, additional weight measurements would be required for model calibration. Thus, the benefits of the statistical approach are that it is less costly and its reliability is known, but its use is dependent on data collected for the variable of interest (weight, in this case) under the circumstances of interest. The mechanistic approach has wider application and a clear rationality (the total

analytical problem. More complex modeling will not necessarily assure that uncertainty is reduced, and in fact can compound problems of uncertain predictions. As discussed below, accounting for uncertainty and representing watershed processes are two of the possible criteria that need to be considered when selecting an analytical model for TMDL development.

TMDLs, which are typically evaluated through predictive modeling, lead to decisions concerning controls on pollutant sources or other stressors. Thus, models used in TMDL analysis provide "decision support."

equals the sum of the parts), but it requires more time and effort, and, unless some data are collected for the variable of interest under similar circumstances, its error characteristics are unknown.

Of course, in practice, mechanistic and statistical modelers often make considerable use of each other's techniques. In the classroom analogy, for example, it would make sense for the statistically inclined teacher to make more detailed measurements of the weighed students' dimensions and develop a multivariate regression model of weight as a function of torso volume, leg volume, etc., rather than height alone. The more complex model could be applied to a wider range of body builds. Moreover, the regression coefficients would represent the estimated densities of different parts of the body. These could be compared with the textbook values of body density as a test of the rationality of the model. Conversely, the mechanistic teacher might use body density data from the textbook to adjust the height–weight regression equations for use with different age and ethnic groups. This would eliminate the need for collecting additional weight data for these groups.

It is also worth distinguishing a third type of model termed *stochastic* that is widely used in engineering applications and that may have a useful role in TMDL modeling. The objective of stochastic modeling is to simulate the statistical behavior of a system by imposing random variability on one or more terms in the model. Such models are usually fundamentally mechanistic, but avoid mechanistic description of complex processes by using simpler randomized terms. Stochastic models generally require a large number of measurements of certain variables (e.g., inputs, state variables) in order to correctly characterize their random behavior. As an example, consider a mechanistic model of river water quality that includes randomly generated streamflow and pollutant loads. If the randomly generated inputs are realistic (both individually and in relation to each other), then the output may provide a very useful description of the variability to expect in the water quality of the river.

Box 4-2 lists *desirable* model selection/evaluation criteria in consideration of the decision support role of models in the TMDL process. The list is intended to characterize an ideal model. Given the limitations of existing models, it should not be viewed as a required checklist for attributes that all present-day TMDL models must have.

EPA has supported water quality model development for many years and, along with the U.S. Geological Survey (USGS), the U.S. Army Corps of Engineers, and the U.S. Department of Agriculture, is responsi-

BOX 4-2
Model Selection Criteria

A predictive model should be broadly defined to include both mathematical expressions and expert scientific judgment. A predictive model useful for TMDL decision support ideally should have the following characteristics:

1. *The model focuses on the water quality standard.* The model is designed to quantitatively link management options to meaningful response variables. This means that it is desirable to define the TMDL endpoints (e.g., pollutant sources and standard violation parameter) and incorporate the entire "chain" from stressors to response into the modeling analysis. This also means that the spatial/temporal scales of the problem and the model should be compatible.

2. *The model is consistent with scientific theory.* The model does not err in process characterization. Note that this is different from the often-stated goal that the model correctly represents processes, which, for terrestrial and aquatic ecosystems, cannot be achieved.

3. *Model prediction uncertainty is reported.* Given the reality of prediction errors, it makes sense to explicitly acknowledge the prediction uncertainty for various management options. This provides decision-makers with an understanding of the risks of options, and allows them to factor this understanding into their decisions. To do this, prediction error estimates are required.

4. *The model is appropriate to the complexity of the situation.* Simple water quality problems can be addressed with simple models. Complex water quality problems may or may not require the use of complex models (as discussed later in this chapter and in Chapter 5).

5. *The model is consistent with the amount of data available.* Models requiring large amounts of monitoring data should not be used in situations where such data are unavailable.

6. *The model results are credible to stakeholders.* Given the increasing role of stakeholders in the TMDL process, it may be necessary for modelers to provide more than a cursory explanation of the predictive model.

7. *Cost for annual model support is an acceptable long-term expense.* Given growth and change, water quality management will not end with the initial TMDL determination. The cost of maintaining and updating the model must be tolerable over the long term.

8. *The model is flexible enough to allow updates and improvements.* Research can be expected to improve scientific understanding, leading to refinements in models.

ble for most models currently being applied for TMDL development. Agency-wide, EPA has funded model development and technology transfer activities for a wide range of models. The greatest concentration of this effort has been at the Center for Exposure Assessment Modeling (CEAM). In contrast to the broad perspective found within EPA as a whole, CEAM has demonstrated a clear preference for mechanistic models, as evidenced by their adoption of the BASINS modeling system (EPA, 2001) as the primary TMDL modeling framework.

Models developed at the CEAM and incorporated into BASINS place high priority on correctly describing key processes, which is related to but different from model selection criterion #2 (see Box 4-2). It is important to recognize that placing priority on ultimate process description often will come at the expense of the other model selection criteria. For one thing, an emphasis on process description tends to favor complex mechanistic models over simpler mechanistic or empirical models and may result in analyses that are more costly than is necessary for effective decision-making. In addition, physical, chemical, and biological processes in terrestrial and aquatic environments are far too complex to be conceptually understood or fully represented in even the most complicated models. For the purposes of the TMDL program, the primary purpose of modeling should be to support decision-making. Our inability to completely describe all relevant processes can be accounted for by quantifying the uncertainty in the model predictions.

UNCERTAINTY ANALYSIS IN WATER QUALITY MODELS

The TMDL program currently accounts for the uncertainty embedded in the modeling exercise by applying a margin of safety (MOS). As discussed in Chapter 1, the TMDL can be represented by the following equation:

$$TMDL = \Sigma WLA + \Sigma LA + MOS$$

This states that the TMDL is the sum of the present and near future load of pollutants from point sources and nonpoint and background sources to receiving waterbodies plus an adequate margin of safety (MOS) needed to attain water quality standards.

One possible metric for the point source waste load allocation (ΣWLA) and the nonpoint source load allocation (ΣLA) is mass per unit time, where time is expressed in days. However, other units of time may actually be more appropriate. For example, it may be better to use a season as the time unit when the TMDL is calculated for lakes and reservoirs, or a year when contaminated sediments are the main stressor.

EPA (1999) gives additional ways in which a TMDL can be expressed:

- the required reduction in percentage of the current pollution load to attain and maintain water quality standards,
- the required reduction of pollutant load to attain and maintain riparian, biological, channel, or morphological measures so that water quality standards are attained and maintained, or
- the pollutant load or reduction of pollutant load that results from modifying a characteristic of a waterbody (e.g., riparian, biological, channel, geomorphologic, or chemical characteristics) so that water quality standards are attained and maintained.

The MOS is sometimes a controversial component of the TMDL equation because it is meant to protect against potential water quality standard violations, but does so at the expense of possibly unnecessary pollution controls. Because of the natural variability in water quality parameters and the limits of predictability, a small MOS may result in nonattainment of the water quality goal; however, a large MOS may be inefficient and costly. The MOS *should* account for uncertainties in the data that were used for water quality assessment and for the variability of background (natural) water quality contributions. It should also reflect the reliability of the models used for estimating load capacity.

Under current practice, the MOS is typically an arbitrarily selected numeric safety factor. In other cases, a numeric value is not stated, and rather conservative choices are made about the models used and the effectiveness of best management practices. Consistent with our concerns, NRC (2000) notes that since parameters involved in the TMDL determination are probabilistic and the MOS is a measure of uncertainty, the MOS should be determined through a formal uncertainty and error propagation analysis. There is also a compelling practical reason for explicit and thorough quantification of uncertainty in the TMDL via the MOS—reduction of the MOS can potentially lead to a significant reduction in TMDL implementation cost. On this basis alone, EPA should place a high priority on estimating TMDL forecast uncertainty and on selecting and developing TMDL models with minimal forecast error.

Model prediction error can be assessed in two ways. First, Monte Carlo simulation can be used to estimate the effect of model parameter error, model equation error, and initial/boundary condition error on prediction error. This process is data-intensive and may be computationally unwieldy for large models. A second and simpler alternative is to compare predictions with observations, although the correct interpretation of this analysis is not as straightforward as it may seem. If a model is "overfitted" to calibration data and the test or "verification" data are not substantially different from the calibration data, the prediction–observation comparison will underestimate the prediction error. The best way to avoid this is to obtain independent verification data substantiated with a statistical comparison between calibration data and verification data.

To date, we are aware of no thorough error propagation studies with the mechanistic models favored by EPA (by thorough, we mean that all errors and error covariance terms are estimated and are plausible for the application). Further, the track record associated with even limited uncertainty analyses is not encouraging for water quality models in general. Among empirical models, only the relatively simple steady-state nutrient input–output models have undergone reasonably thorough error analyses. For example, Reckhow and Chapra (1979) and Reckhow et al. (1992) report prediction error of approximately 30 percent to 40 percent for cross-system models that predict average growing season total phosphorus or total nitrogen concentration based on measured annual loading. Prediction errors are likely to be higher for applications based on estimated or predicted loading. Prediction error will be higher still when these simple models are linked to statistical models to predict chlorophyll *a*, Secchi disk transparency, or an integrative measure of biological endpoints.

Most error analyses conducted on mechanistic water quality models have also focused on eutrophication, so relatively little is known of prediction error for toxic pollutants, microorganisms, or other important stressors. In one of the few relatively thorough error propagation studies, Di Toro and van Straten (1979) and van Straten (1983) used maximum likelihood to determine point estimates and covariances for parameters in a seasonal phytoplankton model for Lake Ontario. Of particular note, they found that prediction error decreased substantially when parameter covariances were included in error propagation, underscoring the importance of including covariance terms in error analyses. This result occurred because, while individual parameters might be highly uncertain, specific *pairs* of parameters (e.g., the half saturation constant and the maximum growth rate in the Michaelis–Menten model) may vary in a predictable way (expressed through covariance) and thus may be *collectively* less uncertain. Di Toro and van Straten found the prediction coefficient of variation to range from 8 percent (for nitrate-N) to 390 percent (for ammonia-N), with half of the values falling between 44 percent and 91 percent. Zooplankton prediction errors tended to be much higher. Beck (1987) found that the error levels cited in these studies are typical of those reported elsewhere. There is evidence to suggest that the current models of water quality, in particular, the larger models, are capable of generating predictions to which little confidence can be attached (Beck, 1987).

The need for understanding the prediction uncertainty of chosen models is not new. Indeed, recent TMDL modeling and assessment guidance from EPA often mentions the importance of formal uncertainty analysis in determining the MOS (EPA, 1999). However, EPA has consistently failed to either recommend predictive models that are amenable to thorough uncertainty analysis or provide adequate technical guidance for reliable estimation of prediction error.

Conclusions and Recommendations

1. EPA needs to provide guidance on model application so that thorough uncertainty analyses will become a standard component of TMDL studies. Prediction uncertainty should be estimated in a rigorous way, and models should be evaluated and selected considering the prediction error need. The limited error analysis conducted within the QUAL2E-UNCAS model (Brown and Barnwell, 1987) was a start, but there has been little progress at EPA in the intervening 14 years.

2. The TMDL program currently accounts for the uncertainty embedded in the modeling exercise by applying a margin of safety (MOS); EPA should end the practice of arbitrary selection of the MOS and instead require uncertainty analysis as the basis for MOS determination. Because reduction of the MOS can potentially lead to a significant reduction in TMDL implementation cost, EPA should place a high priority on selecting and developing TMDL models with minimal forecast error.

3. Given the computational difficulties with error propagation for large models, EPA should selectively target some postimplementation TMDL compliance monitoring for verification data collection to assess model prediction error. TMDL model choice is currently hampered by the fact that relatively few models have undergone thorough uncertainty analysis. Postimplementation monitoring at selected sites can yield valuable data sets to assess the ability of models to reliably forecast response. Large or complex models that pose an overwhelming computational burden for Monte Carlo simulation are particularly good candidates for this assessment.

MODELS FOR BIOTIC RESPONSE: A CRITICAL GAP

The development of models that link stressors (such as chemical pollutants, changes in land use, or hydrologic alterations) to biological responses is a significant challenge to the use of biocriteria and for the TMDL program. There are currently no protocols for identifying stressor reductions necessary to achieve certain biocriteria. A December 2000 EPA document (EPA, 2000) on relating stressors to biological condition suggests how to use professional judgment to determine these relationships, but it offers no other approaches. As discussed below, informed judgment can be effectively used in simple TMDL circumstances, but in more complex systems, empirical or mechanistic models may be required.

There have been some developments in modeling biological responses as a function of chemical water quality. One approach attempts to describe the aquatic ecosystem as a mechanistic model that includes the full sequence of processes linking biological conditions to pollutant sources; this typically results in a relatively complex model and depends heavily on scientific knowledge of the processes. The alternative is to build a simpler empirical model of a single biological criterion as a function of biological, chemical, and physical stressors. Both approaches have been pursued in research dating back at least 30 years, and there has been some progress on both fronts. One promising recent approach is to combine elements of each of these methods. For example, Box 4-3 describes a probability network model that has both mechanistic and empirical elements with meaningful biological endpoints.

Advances in mechanistic modeling of aquatic ecosystems have occurred primarily in the form of greater process (especially trophic) detail and complexity, as well as in dynamic simulation of the system (Chapra, 1996). Still, mechanistic ecosystem models have not advanced to the point of being able to predict community structure or biotic integrity. Moreover, the high level of complexity that has been achieved with this

> **BOX 4-3**
>
> **Neuse Estuary TMDL Modeling**
>
> The Neuse Estuary is listed for chlorophyll a violations (exceedances of 40 ˜g /l), and nitrogen is the pollutant for which a TMDL is developed. Two distinct estuarine models have been developed to guide the TMDL process; one is a two-dimensional process model (CE-Qual-W2), and the other is a probability (Bayes) network model (Borsuk, 2001) depicted in Figure 1.
> This probability network model has several appealing features that are compatible with the modeling framework proposed here:
> - The probabilities in the model are an expression of uncertainty.
> - The conditional probabilities characterizing the relationships described in Figure 1 reflect a combination of simple mechanisms, statistical (regression) fitting, and expert judgment.
> - Some of the model endpoints—estimated using judgmental probability elicitation, which is a rigorous, established process for quantifying scientific knowledge (Morgan and Henrion, 1990)—such as "shellfish survival" and "number of fishkills," characterize biological responses that are more directly meaningful to stakeholders and can easily be related to designated use.
>
> The Neuse Bayes network is a waterbody model; it is being linked to the USGS SPARROW watershed model for allocation of the TMDL.
>
> *Continues*

approach has made it difficult to use statistically rigorous calibration methods and to conduct comprehensive error analyses (Di Toro and van Straten, 1983; Beck, 1987).

The empirical approach depends on a statistical equation in which the biocriterion is estimated as a function of a stressor variable. Success with this empirical approach has been primarily limited to models of relatively simple biological metrics such as chlorophyll *a* (Peters, 1991; Reckhow et al., 1992). For reasons that are not entirely clear, empirical models of higher-level biological variables, such as indices of biotic integrity, have not been widely used. Regressions of biotic condition on chemical water quality measures are potentially of great value in TMDL development because of their simplicity and transparent error characteristics. Two accuracy issues, however, need to be considered. First is the obvious question of whether the level of statistical correlation between biotic metrics and pollutant concentrations is strong enough that prediction errors will be acceptable to regulators and stakeholders. A second

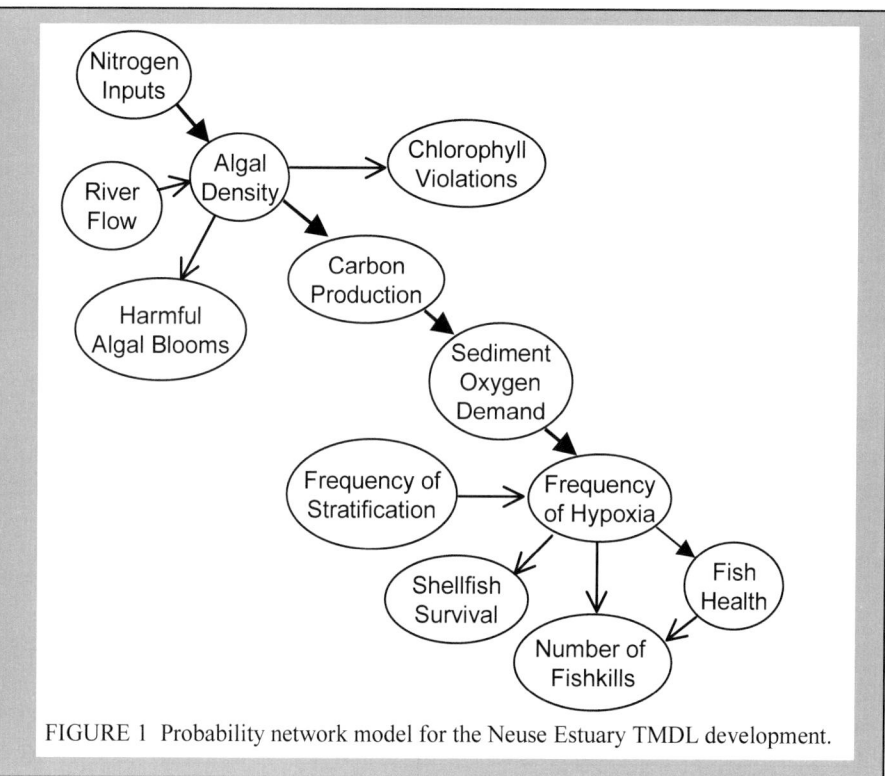

FIGURE 1 Probability network model for the Neuse Estuary TMDL development.

and more difficult issue is that of gaining assurance of a cause–effect relationship between chemical predictors and biotic metrics. The construction of empirical models of biotic condition would benefit greatly from (1) observational data that show the effects of changes in chemical concentrations over a time period when other factors have remained relatively constant and (2) inclusion of as many factors that are relevant to biotic condition as possible. The latter, of course, increases the requirement for observational data. Despite these limitations, in the near term, empirical models may more easily fill the need for biological response models than would mechanistic models.

Conclusions and Recommendations

1. EPA should promote the development of models that can more effectively link environmental stressors (and control actions) to biological responses. Both mechanistic and empirical models should be explored, although empirical models are more likely to fill short-term needs. Such models are needed to promote the wider use of biocriteria at the state level, which is desirable because biocriteria are a better indicator of designated uses than are chemical criteria.

ADDITIONAL MODEL SELECTION ISSUES

Data Required

The use of complex mechanistic models in the TMDL program is warranted if it helps promote the understanding of complex systems, as long as uncertainties in the results are reported and incorporated into decision-making. However, there may be a tendency to use complex mechanistic models to conduct water quality assessments in situations with little useful water quality data and/or involving major remediation expenditures or legal actions. In these situations, there is usually a common belief that the expected realism in the model can compensate for a lack of data, and the complexity of the model gives the impression of credibility. However, given that uncertainty in models is likely to be exacerbated by a lack of data, the recommended strategy is to begin with a simple modeling study and iteratively expand the analysis as needs and new information dictate.

For example, a simple analysis using models like those described by EPA (Mills et al., 1985) as screening procedures could be run quickly at low cost to begin to understand the issues. This understanding might suggest (perhaps through sensitivity analysis) that data should be collected on current land use, or that a limited monitoring program is warranted. Following acquisition of that information/data, a revised (perhaps more detailed) model could be developed. This might result in the TMDL (to be further evaluated using adaptive implementation as described in Chapter 5), or it might lead to further data collection and refinement of the model. This strategy for data-poor situations makes efficient use of resources and targets the effort toward information and models that will reduce the uncertainty as the analysis proceeds.

The data required for TMDL model development will be a function of the water quality criterion and its location and the analytical procedures used to relate the stressors to the criterion. Data needs may include hydrology (streamflow, precipitation), ambient water quality measures, and land use and elevation in a watershed (see Box 4-4 for more infor-

BOX 4-4
Data Requirements for TMDL Modeling of Pollutants

The data and information required for TMDL modeling must reflect the parameters that affect attainability of water quality standards. Many of the models used today have extremely large data requirements, a fact that must be addressed prior to TMDL development so that adequate data collection can occur.

Flow Data. Critical to the process of calibrating and verifying models are flow data, from sources and various locations in the receiving water. Flow data are generally high in quality if gathered as part of unidirectional stream surveys, but become less reliable in areas subject to tidal effects. The USGS is generally considered to be the most reliable source for long-term, high-quality data sets. Tidal records are available, historically and for predictive purposes, for many coastal waters in the United States from the National Oceanic and Atmospheric Administration. Some states have maintained long-term gages in coastal waters, but these are usually few in number.

Ambient Water Quality Data. A number of federal agencies, state agencies, regional organizations, and research groups collect surface water quality data. Many of these data are retrievable over the Internet, particularly data from the USGS and EPA. Although there is no universal repository for all surface water quality data, the STORET database is the most comprehensive. Because methods of collection and analysis may vary, there is a need for QA/QC of these data.

Land Use Data. All states should have access to a series of land use records and projections. For ease of use, the land use data sets should be made available as Geographic Information System (GIS) coverages. EPA has provided default coverages as a component of its BASINS model. For TMDL purposes, land use data are required for the time period over which water quality data are available in order to calibrate and validate models. Projected land use data are needed for predicting future scenarios. The overall quality of these land use data will vary, often as a function of the level of ground-truthing that was done or the accuracy of the predictions for future land use changes.

Point Source Data. Model inputs may include measured values of pollutant loading from point sources (e.g., based on information reported on NPDES Discharge Monitoring Reports submitted by permitted facilities). Other possible data sources include results from periodic compliance inspections and wasteload allocation studies, or data collected as part of field surveys done in support of the TMDL. Such data are generally available and reliable.

continues

BOX 4-4 Continued

Nonpoint Source Data. Data on pollutant loadings from nonpoint sources are much less available and reliable than data from point sources. This is partly because during high-flow, high-rainfall events, monitoring is only infrequently conducted. For nonpoint sources, Event Mean Concentrations (EMCs) are needed to estimate the loadings that are delivered from each significant land use in a basin. EMCs are useful tools in providing estimated nonpoint source loads. Given the wide range of actual loads that may be associated with nonpoint sources, these estimates frequently represent the best science available.

Atmospheric Deposition. Data on pollutant loadings from atmospheric deposition have been compiled by the National Atmospheric Deposition Program/National Trends Network (NADP/NTN) using a nationwide network of precipitation-monitoring sites to generate reliable estimates of loads for many parameters. However, unlike watersheds, airsheds vary in size, depending upon the pollutant of concern and its specific forms and chemistry. Assessing the atmospheric contribution to any one basin is complicated by variations attributable to factors such as seasonal shifts in prevailing winds and distance from contributing sources. Thus, it is currently difficult to differentiate impacts from local sources vs. remote sources. For example, although significant work has been done in the northeastern United States to link sources of nitrous oxides with the areas subject to impact, similar studies elsewhere are not routinely available. Data for parameters other than those covered by NADP sites, as well as data on basin-specific wet and dry atmospheric deposition rates, are also scant.

Legacy/Upstream Sources. For many impaired waters, states will need to identify and estimate loads attributed to legacy sources (e.g., PCBs, DDT, or the phosphorus-laden lake sediments) and upstream sources (those entering a waterbody segment upstream of the watershed currently being studied). The availability and reliability of such data vary widely across the nation.

Best Management Practices. TMDL development will in many cases require estimates of the treatment efficiency for a best management practice (BMP). Such data are generally not available, except for a small number of well-studied stormwater BMPs and a limited number of pollutants (see NRC, 2000). To account for these deficiencies, states might use best professional judgment to estimate the percent reduction, taking into account treatment provided by similar BMPs and stakeholder input. EPA has recently provided funding for a national database designed to help states track the effectiveness of BMPs as they are developed and evaluated. Databases of BMP effectiveness are currently available at ASCE (1999) and Winer (2000).

mation). TMDL development will also likely require data on point/ nonpoint sources and pollutant loads, atmospheric deposition, the effectiveness of current best management practices, and legacy/upstream pollutant sources. Because the amount of available data varies with site, there is no absolute minimum data requirement that can be universally set for TMDL development. Data availability is one source of uncertainty in the development of models for decision support. Although there are other sources of uncertainty as well, models should be selected (simple vs. complex) in part based on the data available to support their use.

Simple vs. Complex Models

The model selection criteria concerning cost, flexibility, adaptability, and ease of understanding (Box 4-2) all tend to favor simple models, although they may fail to adequately satisfy the first criterion. There are many situations, however, when an exceedingly simple model is all that is needed for TMDL development, particularly when combined with adaptive implementation (to be discussed in Chapter 5). For example, it is not uncommon in many states for farm fields to straddle small streams, with cows being allowed to freely graze in and around the stream. If a downstream water quality standard is violated, a simple mental model linking the cows to the violation, and subsequent actions in which the first step might be to limit cow access to the riparian corridor, may ultimately be sufficient for addressing the impairment. This example is certainly not intended to suggest that all TMDLs will be simple, but it does suggest the value of simple analyses and iterative implementation. Box 4-5 presents a relatively simple modeling exercise (based on a statistical rather than mechanistic model) that was used successfully to develop a TMDL for clean sediment.

With regard to mechanistic models, there is no intrinsic reason to choose the particular scales that have become the basis for representing processes in the majority of mechanistic water quality models. As an alternative, Borsuk et al. (2001) have shown that it is possible to specify relatively simple mechanistic descriptions of key processes in aquatic ecosystems, which limits the dimension of the parameter space so that parameters may be estimated using least squares or Bayesian methods on the available data. The SPARROW model (Smith et al., 1997) is another more statistically based alternative that includes terms and functions that reflect processes. These efforts suggest that a fruitful research direction for the TMDL program is the development of models that are based on

**BOX 4-5
Use of a Simple Empirical Model:
Suspended Sediment Rating Curve for Deep Creek, MT**

One relatively simple form of model that has been used successfully in many TMDL applications is a statistical regression of a water quality indicator on one or more predictor variables. The indicator may be either the pollutant named in the TMDL or a related metric used to determine impairment but not directly involved in the TMDL analysis. Such a model was used to develop a TMDL for suspended sediment in Deep Creek, MT (see Endicott, 1996). The designated use of that waterbody was to support a cold water fishery and its associated biota, especially to provide high-quality spawning areas to rainbow and brown trout from a nearby reservoir. The reservoir and the river provide a blue-ribbon trout fishery. Analyzing the effects of suspended sediment on salmonids is complicated by the fact that sediment concentrations in western trout streams increase dramatically with streamflow in healthy as well as sediment-impaired streams, but are lower at any given flow in the healthy streams than in the impaired streams. Suspended sediment concentrations at all stages of the hydrograph are important biologically.

To develop a sediment TMDL at this site, modelers compared the relationship of sediment concentration to streamflow (known as the "sediment rating curve") at the impaired site to the corresponding sediment rating curve for an unimpaired reference site. Rating curves were developed by regressing sediment concentration on streamflow. In the case of Deep Creek, the sediment–flow relationship is approximately linear with a slope of 0.51 mg l^{-1} per $ft^3 sec^{-1}$. Based on rating curves for reference streams of similar size in the area (Endicott, 1996), an appropriate slope would be 0.26 mg l^{-1} per $ft^3 sec^{-1}$. Thus, the goal of TMDL implementation is to lower the Deep Creek ratio by about half. According to the approved TMDL management plan, certain channel modifications and a combination of riparian and grazing BMPs are expected to reduce the slope of the sediment rating curve and restore the health of the trout fishery. Determination of whether the control measures have reduced the rating curve slope to the target level can be accomplished in the future by a hypothesis test on the slope parameter of the revised regression of concentration on flow. The Type 1 and Type 2 error rates for this decision-making method will relate directly to the statistical confidence limits on the estimated slope parameter, and are controllable through the quantity of monitoring data collected after the control measures are in place.

There are several aspects of this modeling approach that make it well suited to the TMDL problem. The analysis was simple to carry out and relatively easy for stakeholders to understand. Despite its simplicity, the model focuses on a critical aspect of the Deep Creek ecosystem—suspended sediment concentrations over the entire hydrograph. Future decision-making on the success of the management plan can be based on an objective test with known error rates that are controllable through monitoring.

process understanding yet are fitted using statistical methods on the observational data.

Pilot Watersheds

Another approach to consolidate modeling efforts and develop TMDLs more efficiently is the pilot watershed concept[14]. Many TMDLs involve small- to medium-sized watersheds that have a dominating non-point source pollution problem (e.g., the Corn Belt region, watersheds draining forested areas, or suburban watersheds). Watersheds located in the same ecoregion may have similar water quality problems and solutions. Thus, a detailed modeling study of one or two benchmark watersheds can provide problem identification and solutions. These findings could potentially be extrapolated to less investigated but similar watersheds.

Conclusions and Recommendations

If accompanied by uncertainty analysis, many existing models can be used to develop TMDLs in an adaptive implementation framework. Adaptive implementation, discussed in detail in Chapter 5, will allow for both model development over time and the use of currently available data and methods. It provides a level of assurance that the TMDL will ultimately be successful even with high initial forecast uncertainty.

1. EPA should not advocate detailed mechanistic models for TMDL development in data-poor situations. Either simpler, possibly judgmental, models should be used or, preferably, data needs should be anticipated so that these situations are avoided. The strategy of accounting for data-limited TMDLs with increasingly detailed modelse

---

[14] In various forms, "pilot watersheds" have for years been the basis for understanding land use impacts on water quality. The concept is implicit in the acceptance and use of export coefficients for pollutant load assessment. A prominent example is the series of PLUARG (Pollution from Land Use Activities-Reference Group) studies to determine the total loads of pollutants to the Great Lakes. The group used several pilot watersheds on each side of the border and extrapolated the detailed monitoring and modeling results into the entire Great Lakes basin.

needs rigorous verification before it should be endorsed and implemented. Starting with simple analyses and iteratively expanding data collection and modeling as the need arises is the best approach.

2. EPA needs to provide guidance for determining the level of detail required in TMDL modeling that is appropriate to the needs of the wide range of TMDLs to be performed. The focus on detailed mechanistic models has resulted in complex, costly, time-consuming modeling exercises for single TMDLs, potentially taking away resources from hundreds of other required TMDLs. Given the variety of existing watershed and water quality models available, and the range of relevant model selection criteria, EPA should expand its focus beyond mechanistic process models to include simpler models. This will support the use of adaptive implementation.

3. EPA should support research in the development of simpler mechanistic models that can be fully parameterized from the available data. This would lead to models that meet several model selection criteria present in Box 4-2, such as consistency with theory, assessing uncertainty, and consistency with available data.

4. To more efficiently use scarce resources, EPA should approve the use of pilot watersheds for TMDL modeling. Rather than detailed models being prepared for every impaired waterbody, pilot TMDLs could be prepared in detail for a benchmark watershed (e.g., a typical suburban or agricultural watershed), and the results could be extrapolated to similar watersheds located in the same ecoregion. The notion of extending modeling results to similar areas, which underlies the present-day use of export coefficients, is reasonable if applied in the framework of adaptive implementation. Such a framework, coupled with the rapid application of specific controls/approaches in a number of watersheds, can reveal where techniques do or do not work and can allow for appropriate modifications.

REFERENCES

ASCE. 1999. National Stormwater Best Management Practices (BMP) Database. Version 1.0. Prepared by Urban Water Resources Research Council of ASCE, and Wright Water Engineers, Inc., Urban Drainage and Flood Con-

trol District, and URS Greiner Woodward Clyde, in cooperation with EPA Office of Water, Washington, DC. User's Guide and CD.
Beck, M. B. 1987. Water quality modeling: a review of the analysis of uncertainty. Water Resources Research 23:1393–1442.
Beven, K. J. 1996. A discussion of distributed hydrological modeling. Distributed hydrological modeling. M. B. Abbott and J. C. Refsgaard, Ed. Dordrecht, Netherlands: Kluwer Academic Publishers. pp. 255–278.
Borsuk, M. E. 2001. A Probability (Bayes) Network Model for the Neuse Estuary. Unpublished Ph.D. dissertation. Duke University.
Borsuk, M. E., C. A. Stow, D. Higdon, and K. H. Reckhow. 2001. A Bayesian hierarchical model to predict benthic oxygen demand from organic matter loading in estuaries and coastal zones. Ecological Modeling (In press).
Brown, L. C., and T. O. Barnwell, Jr. 1987. The enhanced stream water quality models QUAL2E and QUAL2E-UNCAS: documentation and user manual. EPA-600/3-87/007. Athens, GA: EPA Environmental Research Laboratory.
Chapra, S. C. 1996. Surface Water Quality Modeling. New York: McGraw-Hill. 844 p.
Di Toro, D. M., and G. van Straten. 1979. Uncertainty in the Parameters and Predictions of Phytoplankton Models. Working Paper WP-79-27, International Institute for Applied Systems Analysis, Laxenburg, Austria.
Endicott, C. L., and T. E. McMahon. 1996. Development of a TMDL to reduce nonpoint source sediment pollution to Deep Creek, Montana. Report to Montana Department of Environmental Quality, Helena, Montana. Montana State University, Bozeman, Montana.
Environmental Protection Agency (EPA). 1994. Water Quality Standards Handbook: Second Edition. EPA 823-B-94-005a. Washington, DC: EPA Office of Water.
EPA. 1999. Draft Guidance for water Quality-based Decisions: The TMDL Process (Second Edition), Washington, DC: EPA Office of Water.
EPA. 2000. Stressor Identification Guidance Document. EPA-822-B-00-025. Washington, DC: EPA Office of Water and Office of Research and Development.
EPA. 2001. BASINS Version 3.0 User's Manual. EPA-823-B-01-001. Washington, DC: EPA Office of Water and Office of Science and Technology. 337p.
Mills, W. B., D. B. Porcella, M. J. Ungs, S. A. Gherini, K. V. Summers, L. Mok, G. L. Rupp, G. L. Bowie, and D. A. Haith. 1985. Water Quality Assessment: A Screening Procedure for Toxic and Conventional Pollutants in Surface and Ground Water, Parts I and II. EPA/600/6-85/002a,b.
Morgan, M. G., and M. Henrion. 1990. Uncertainty. New York: Cambridge University Press. 332 p.
National Research Council (NRC). 2000. Watershed Management for Potable Water Supply—Assessing the New York City Strategy. Washington, DC: National Academy Press.

Peters, R. H. 1991. A critique for ecology. Cambridge: Cambridge University Press. 366 p.

Reckhow, K. H., and Chapra, S. C. 1979. Error analysis for a phosphorus retention model. Water Resources Research 15:1643–1646.

Reckhow, K. H., Coffey, S. C., Henning, M. H., Smith, K. and Banting, R. 1992. Eutromod: Technical Guidance and Spreadsheet Models for Nutrient Loading and Lake Eutrophication. Duke University School of the Environment, Durham, NC.

Smith, R. A., G. E. Schwarz, and R. B. Alexander. 1997. Regional interpretation of water-quality monitoring data. Water Resources Research 33(12):2781–2798.

Spear, R., and G. M. Hornberger. 1980. Eutrophication in Peel Inlet – II. Identification of critical uncertainties via generalized sensitivity analysis. Water Research 14:43–49.

Ulanowicz, R. E. 1997. Ecology, the ascendant perspective. New York: Columbia University Press. 201p.

van Straten, G. 1983. Maximum likelihood estimation of parameters and uncertainty in phytoplankton models. In: M. B. Beck and G. van Straten (Editors), Uncertainty and Forecasting of Water Quality. Berlin: Springer Verlag.

Winer, R. 2000. National Pollutant Removal Performance Database for Stormwater Treatment Practices, Second Edition. Center for Watershed Protection, Ellicott City, MD. Prepared for EPA Office of Science and Technology, in association with Tetra Tech, Fairfax VA.

5
Adaptive Implementation for Impaired Waters

Water quality assessment is a continuous process. The finding of an impaired waterbody during assessment triggers a sequence of events that may include listing of the water, development of a Total Maximum Daily Load (TMDL), planning of state and federal actions, and implementation events designed to comply with water quality standards—all of which are characterized by uncertainty. This chapter describes the process of adaptive implementation of a water quality plan. Adaptive implementation simultaneously makes progress toward achieving water quality standards while relying on monitoring and experimentation to reduce uncertainty.

SCIENCE AND THE TMDL PROCESS

The planning sequence of moving from data to analysis to information and knowledge is supposed to provide confidence that the sometimes costly actions to address a water quality problem are justified. A desire for this confidence is often behind the call for "sound science" in the TMDL program. However, the ultimate way to improve the scientific foundation of the TMDL program is to incorporate the *scientific method*, not simply the results from analysis of particular data sets or models, into TMDL planning. The scientific method starts with limited data and information from which a tentatively held hypothesis about cause and effect is formed. The hypothesis is tested, and new understanding and new hypotheses can be stated and tested. By definition, science is this process of continuing inquiry. Thus, calls to make policy decisions based on the "the science," or calls to wait until "the science is complete," reflect a misunderstanding of science. Decisions to pursue some actions must be made, based on a preponderance of evidence, but there may be a need to continue to apply science as a process (data collection and tools of analysis) in order to minimize the likelihood of future errors.

Many debates in the TMDL community have centered on the use of "phased" and "iterative" TMDLs. Because these terms have particular meanings, this report uses a more general term—adaptive implementation. Adaptive implementation is, in fact, the application of the scientific method to decision-making. It is a process of taking actions of limited scope commensurate with available data and information to continuously improve our understanding of a problem and its solutions, while at the same time making progress toward attaining a water quality standard. Plans for future regulatory rules and public spending should be tentative commitments subject to revision as we learn how the system responds to actions taken early on.

Like other chapters, this chapter discusses a framework for water quality management (shown in Figure 5-1, which is the same as Figure 3-1). Before turning to adaptive implementation, it discusses an important prior step—review of water quality standards. Before a waterbody is placed on the action (303d) list, it is suggested that states conduct a review of the appropriateness of the water quality standard. The standards review may result in the water not being listed as impaired if the standard used for the assessment was found to be inappropriate. On the other hand, the same process may result in a "stricter" standard than was used in the assessment process, in which case the waterbody would have a TMDL plan developed to achieve that revised standard. A review of the water quality standard will assure that extensive planning and implementation actions are directed toward clearly conceived designated uses and associated criteria to measure use attainment.

REVIEW OF WATER QUALITY STANDARDS

Water quality standards are the benchmark for establishing whether a waterbody is impaired; if the standards are flawed (as many are), all subsequent steps in the TMDL process will be affected. Although there is a need to make designated use and criteria decisions on a waterbody and watershed-specific basis, most states have adopted highly general use designations commensurate with the federal statutory definitions. However, an appropriate water quality standard must be defined *before* a TMDL is developed. Within the framework of the Clean Water Act (CWA), there is an opportunity for such analysis, termed use attainability analysis (UAA).

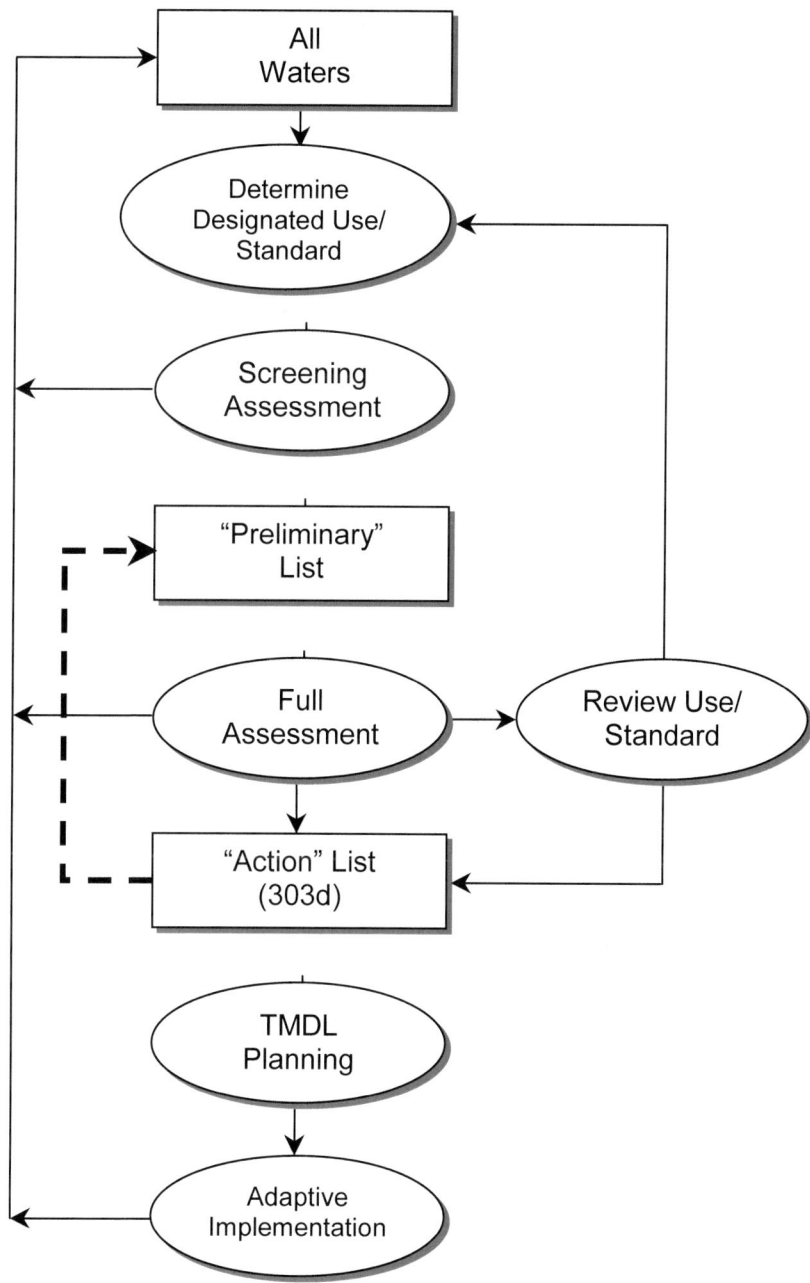

FIGURE 5-1 Framework for water quality management.

A UAA determines if impairment is caused by natural contaminants, nonremovable physical conditions, legacy pollutants, or natural conditions (see Box 5-1). More importantly, a UAA can refine the water quality standard. UAA should result in more stratified and detailed narrative statements of the desired use and measurable criterion. For example, a UAA might refine the designated use and criterion from a statement that the water needs to be fishable to a statement calling for a reproducing trout population. Then one or more criteria for measuring attainment of this designated use are described; these might include minimum dissolved oxygen or maximum suspended sediment requirements. Alternatively, an index to measure biological condition appropriate to the trout fishery designated use, such as an index of biological integrity (IBI), may be defined.

In the 1990s, TMDLs were undertaken for some waterbodies where the designated use was not attainable for reasons that could have been disposed of by a UAA. For example, TMDLs conducted in Louisiana resulted in the conclusion that even implementing zero discharge of a pollutant would not bring attainment of water quality standards (Houck, 1999). A properly conducted UAA would have revealed the true problem—naturally low dissolved oxygen concentrations—before the time and money were spent to develop the TMDL. Unfortunately, UAA has not been widely employed. Novotny et al. (1997) found that 19 states reported no experience with UAA. The majority of states reported a few to less than 100 UAAs, while five states (Indiana, Nebraska, New York, Oklahoma, and Pennsylvania) performed more than 100.

One possible explanation for the failure to widely employ UAA analysis is the absence of useful EPA guidelines. The last technical support manuals were issued in the early 1980s (EPA, 1983) and are limited to physical, chemical and biological analyses. It is presently not clear what technical information constitutes an adequate UAA for making a change to the use designation for a waterbody that will be approved by the EPA.

In addition to being a technical challenge, standards review also has important socioeconomic consequences (see point 6 in Box 5-1). EPA has provided little information on how to conduct socioeconomic analyses or how to incorporate such analyses in the UAA decision. The socioeconomic analysis suggested by EPA is limited to narrowly conceived financial affordability and economy-wide economic impact assessments (e.g., employment effects) (Novotny et al., 1997). However, when setting water quality standards, states may be asked to make decisions in

> **BOX 5-1**
> **Six Reasons for Changing the Water Quality Standard**
>
> The following six situations, which can be revealed by UAA, constitute reasons for changing a designated use or a water quality standard (EPA, 1994). Conducting a UAA does not necessarily preclude the development of a TMDL.
>
> 1. Naturally occurring pollutant concentrations prevent attainment of the use.
> 2. Natural, ephemeral, intermittent, or low flow water levels prevent the attainment of the use unless these conditions may be compensated for by a sufficient volume of effluent discharge without violating state conservation requirements to enable uses to be met.
> 3. Human-caused conditions or sources of pollution prevent the attainment of the use and cannot be remedied or would cause more environmental damage to correct than to leave in place (e.g., as with some legacy pollutants).
> 4. Dams, diversions, or other types of hydrologic modifications preclude the attainment of the use, and it is not feasible to restore the waterbody to its original condition or to operate such modification in a way that would result in the attainment of the use.
> 5. Physical conditions related to the natural features of the waterbody, such as the lack of proper substrate, cover, flow, depth, pools, riffles, and the like, unrelated to water quality, preclude attainment of aquatic life protection uses.
> 6. Controls more stringent that those required by the CWA mandatory controls (Sections 301b and 306) would result in substantial and widespread adverse social and economic impact. This requires developing a TMDL and conducting a socioeconomic impact analysis of the resulting TMDL (Novotny et al., 1997).

consideration of a broader socioeconomic benefit–cost framework than what is currently expected in a UAA. Finally, EPA has offered no guidance on what constitutes an acceptable UAA in waterbodies of different complexity and on what decision criteria will be accepted as a basis for changing a use designation. This is significant because EPA retains the authority to approve state water quality standards. These uncertainties discourage state use of UAA because there is no assurance that EPA will accept the result of the UAA effort as an alternative to a TMDL, especially if the EPA expectation for a UAA will result in significant analytical costs.

Conclusions and Recommendations

1. EPA should issue new guidance on UAA. This should incorporate the following: (1) levels of detail required for UAAs for waterbodies of different size and complexity, (2) broadened socioeconomic evaluation and decision analysis guidelines for states to use during UAA, and (3) the relative responsibilities and authorities of the states and EPA in making use designations for specific waterbodies following a UAA analysis.

2. UAA should be considered for all waterbodies before a TMDL plan is developed. The UAA will assure that before extensive planning and implementation actions are taken, there is clarity about the uses to be secured and the associated criteria to measure use attainment. UAA is especially warranted if the water quality standards used for the assessment were not well stratified. However, the decision to do a UAA for any waterbody should rest with each state.

ADAPTIVE IMPLEMENTATION DESCRIBED

Once a waterbody is on the 303d list, a plan to secure the designated use is developed and a sequence of actions is implemented. The adaptive implementation process begins with initial actions that have a high degree of certainty associated with their water quality outcome. Future actions must be based on (1) continued monitoring of the waterbody to determine how it responds to the actions taken and (2) carefully designed experiments in the watershed. This concurrent process of action and learning is depicted in Figure 5-2.

The plan includes the following related elements: immediate actions, an array of possible long-term actions, success monitoring, and experimentation for model refinement. In choosing *immediate actions*, watershed stakeholders and the state should expect such actions to be undertaken within a fixed time period specified in the plan. If the impairment problem is attributable to a single cause or if the impairment is not severe, then the immediate actions might be proposed as the final solution to the nonattainment problem. However, in more challenging situations, the immediate actions alone should not be expected to completely eliminate the impairment.

Regardless of what immediate actions are taken, there may not be an immediate response in waterbody or biological condition. For example,

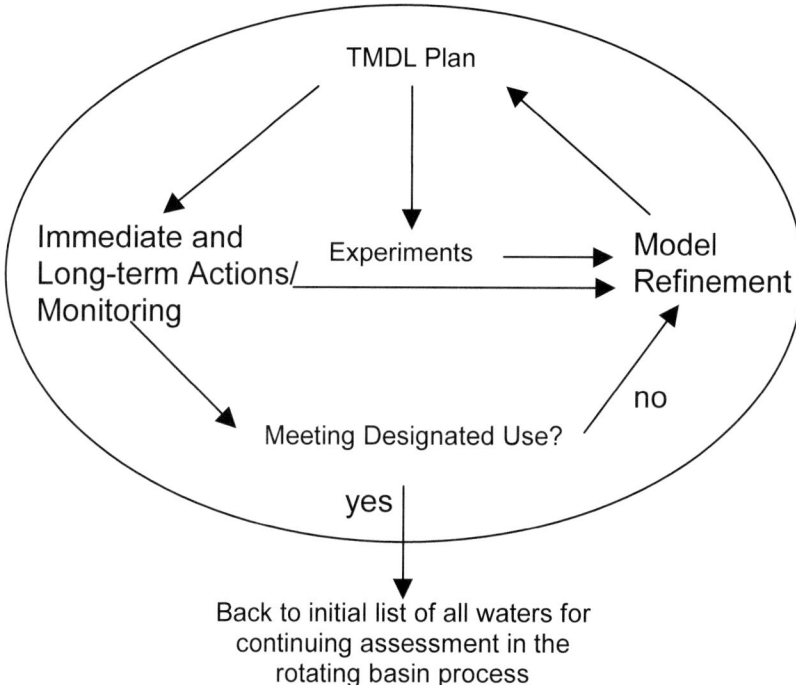

FIGURE 5-2 Adaptive implementation flowchart.

there may be significant time lags between when actions are taken to reduce nutrient loads and resulting changes in nutrient concentrations. This is especially likely if nutrients from past activities are tightly bound to sediments or if nutrient-contaminated groundwater has a long residence time before its release to surface water. For many reasons, lags between actions taken and responses must be expected. As discussed below, the waterbody should be monitored intensively to establish whether the "trajectory" of the measured water quality criterion points toward attainment of the designated use.

Longer-term actions are those that show promise, but need further evaluation and development. They should be formulated in recognition of emerging and innovative strategies for waterbody restoration. The commitment in the plan is to further evaluate such actions based on the collection of additional data, data analysis, and modeling. An adaptive implementation plan would specify analyses of specific long-term alternatives, a schedule for such analyses to be conducted, and a mechanism for supporting such analyses.

Success monitoring follows after implementation actions. If success monitoring shows that the waterbody is meeting water quality standards including designated uses, then no further implementation actions would be taken. Waterbodies should be returned to the "all waters" list (see Figure 5-1) where they will be monitored as a part of the rotating basin process. A primary purpose of success monitoring is to establish compliance with water quality standards and ultimately make the delisting decision. Because state ambient monitoring programs typically have limited resources, it may be necessary to design and implement success monitoring for the TMDL program outside the rotating basin process. Those stakeholders affected by 303d listing and TMDL development may have an incentive to make a significant contribution to the monitoring effort to assure that the water is truly impaired and that the best possible models are being used for plan development. Stakeholder monitoring would be conducted with input on its design by the state.

One of the most important applications of success monitoring data is to revise and improve the initial TMDL forecast over time. This revision of the TMDL model can be formally accomplished using techniques such as Bayesian analysis, data assimilation, or Kalman filtering. For example, a TMDL for total phosphorus, based on a model forecast that included uncertainty analysis, might be implemented to address a chlorophyll *a* standard violation. As part of the implementation program, monitoring would be undertaken to assess success and compliance. At the end of the five-year rotating basin cycle, the original chlorophyll *a* forecast could be combined with the monitoring-based chlorophyll *a* time trajectory to yield a revised forecast of ultimate chlorophyll *a* response. This revised forecast could provide the basis for changes to be implemented during the next five-year cycle in order to meet the water quality standard.

Techniques to accomplish model refinement have existed for some time in a Bayesian context (Reckhow, 1985), and under various labels and modifications, they are being applied in other areas. For example, "data assimilation" (Robinson and Lermusiaux, 2000), a derivative of Bayesian inference, is being widely used in the earth sciences to augment uncertain model forecasts with observations. The Bayesian approach holds particular appeal for adaptive TMDLs because it involves "knowledge updating" that is based on pooling precision-weighted information.

Adaptive Implementation for Impaired Waters 97

The need for *experimentation* to be part of the plan depends on the complexity of the problem and the need to learn more about the system for subsequent model refinement and decision-making. Experiments can, for example, be developed to test the site-specific effectiveness and response time of best management practices (BMPs) (like riparian buffers), to determine the fate and transport of pollutants in runoff, or to answer other questions critical to model refinement. Experiments must be carefully designed and adequately supported (with both funding and staff) to study the effectiveness of actions in the watershed context and to study and learn about watershed processes that are not well understood. TMDL plans for waterbodies with relatively simple problems that can be addressed with high certainty about cause and effect might not include experimentation.

All the actions described above can be used to refine the original TMDL plan so that it better reflects the current state of knowledge about the system and innovative modeling approaches. When revising the TMDL plan, water managers should consider whether the longer-term actions discussed above, or other new alternatives, should be implemented in addition to the immediate actions called for in the original plan. TMDL plans for complicated systems (e.g., a reservoir impacted by multiple nonpoint sources of pollution) can be expected to undergo more revisions before water quality standards (including designated uses) are met than will TMDL plans developed for simple systems.

TMDL IMPLEMENTATION CHALLENGES

Allocation Issues

Plan implementation involves actions taken to reduce all the stressors responsible for the impairment. The allocation of financial and legal responsibility for taking those actions will fall on stakeholders in the watershed, who may not receive public subsidies for taking such actions. Because of these cost consequences, stakeholders want to be sure that water quality standards are appropriate and that total load limits and the limits proposed on other stressors (e.g., flow modifications) are necessary to secure the designated use.

The committee's charge included a request to evaluate the reliability of "the information required to allocate reductions in pollutant loadings among sources." Allocation is *first and foremost a policy decision* on how to distribute costs among different stakeholders in order to achieve a water quality goal. Consider a hypothetical example where three different actions are possible: reduction of pollutant loads from a treatment plant, reductions in pollutant load in runoff from urban areas and farm fields, and increases in stream flow from reduced consumptive irrigation water use. Also suppose that different combinations of all of these actions can achieve the designated use. Allocation becomes a difficult decision because the different combinations will have a different total cost and different levels of perceived fairness. One suggestion might be to choose the combination of actions that minimizes total cost. However, this may result in a cost distribution that places most of the burden on the customers of the treatment plant (for example). An alternative may be to reduce loads from the plants and from runoff by the same proportion; however, this leaves unanswered whether any cost responsibility should fall on the irrigators. Other combinations of actions would have other cost distribution effects.

Although the allocation process is primarily a policy decision, there is one important role that science can play—determining when actions are "equivalent." Water quality management actions are defined to be "equivalent" when their implementation achieves the designated use, taking uncertainty into consideration. Note that there are two aspects of this definition of equivalency. First, equivalency is established with respect to ambient outcomes for the watershed and not in terms of pollutant loading comparisons, which is the way the allocations are described in the standard TMDL equation. Second, the definition recognizes that equivalency must account for the relative uncertainty of different actions with respect to meeting the applicable water quality standard.

One common scenario might be the need to establish equivalency between nitrogen load reductions from a proposed agricultural BMP vs. a proposed wastewater treatment plant improvement. Estimates of the effectiveness of the BMP and wastewater treatment technology can be made in a controlled setting, perhaps with field studies of the BMP and with experiments at the treatment plant. To achieve equivalency, these load reductions must have the same effect on meeting the water quality standard, which would normally be determined using a modeling approach as described in Chapter 4. It is quite possible that the nitrogen load reductions at the sources (the agricultural BMP and the wastewater treatment plant) are different, but they are equivalent in that they are predicted to have an identical effect on the standard. Further, as noted above, equivalency is a function of both the forecasted mean and forecast uncertainty. Thus, if the BMP and wastewater treatment improvement are both forecast to have the same mean effect on the water quality standard, but the wastewater treatment improvement response has less uncertainty, then the actions are not equivalent.

Determining equivalency across sources requires predicting or measuring the results of control actions, rather than simply noting the presence or absence of a particular control technology (the results of which may vary depending on how it is operated and on many other factors). Careful thought must be given to determining meaningful results, especially in those watersheds where actions like flow augmentation or planting of oysters in an estuary are being used as substitutes for, or necessary complements to, load reduction to meet the designated use.

Finally, because it should be focused on water quality outcomes, allocation is dependent on modeling the effects of different actions on waterbody response. Thus, the issues of model selection and uncertainty that were described in Chapter 4 for TMDL development also apply to TMDL allocation. If there is uncertainty about the effect of certain control actions, those who bear the costs may resist taking such actions without further evidence of their worth. Adaptive implementation would support a cautious approach of taking low-cost actions with a high degree of certainty about the outcome, while taking parallel longer-term actions to improve model capabilities and revise control strategies.

Progressing Toward Adaptive Implementation

The TMDL program is limited by an incomplete conceptual understanding of waterbodies and watersheds, by models that are necessarily abstractions from the reality of natural systems, and by limited data for testing hypotheses and/or simulating systems. As a result, it is possible for a waterbody to be identified as impaired when it is not; in such cases, the costs to plan and implement control actions are wasted. On the other hand, it is also possible that an impaired waterbody will not be identified, resulting in other adverse consequences. Many of the stakeholders who addressed the committee expressed concern about the ramifications of uncertainty in the TMDL process. Some cautioned against listing errors, noting that the listing decision can trigger a linear and inflexible process of potentially expensive controls on land use and pollutant discharges that may ultimately prove unwarranted. Others who are concerned that impaired waterbodies will go unidentified advocated more aggressive and comprehensive actions to address problems quickly. These differences in viewpoint can be traced to the policy context that now governs the TMDL program. The committee views adaptive implementation as accommodating this spectrum of opinions.

If adaptive implementation is to be adopted, three policy issues that stand in the way of acceptance of the approach must be addressed. These issues are described without specific recommendations on their solution, except to note that their resolution is needed in order for the TMDL program to fully embrace the scientific method. Criticism of the TMDL program is too often, and sometimes inappropriately, directed at the quality of the data and information, rather than at these underlying policy issues.

1. The listing of a waterbody and the initiation of the TMDL process appear to call for a constraint on total pollutant loading associated with population growth and land use shifts until the designated use is obtained. Given the often weak water quality standards that underlie a listing, the long lag times between actions taken and measured responses, and the uncertainty in our ability to predict what actions will secure a designated use, it is unrealistic to expect that there will be no changes in economic activity and in land uses in a watershed until the designated use has been achieved. A basis for accommodating growth and change in watersheds needs to be established as adaptive implementation proceeds.

2. Many waterbody stressors currently lie outside the CWA regulatory framework, where the only federal enforcement tool available is point source discharge limits. Recognition of this fact was a motivation for EPA's endorsement of the watershed approach in 1991 (EPA, 1993). Nonetheless, in some cases point source permitting is used to impose conditions on point sources that essentially require them to finance control practices for unregulated nonpoint sources (NAPA, 2000). Perceptions of the inequity and the ineffectiveness of such a requirement may be manifested as technical critiques of the TMDL analysis itself. Distributing the cost and regulatory burdens for designated use attainment in a way that is deemed equitable by all stakeholders is critical to future TMDL program success.

3. Watersheds can range in size from a few acres to an area that covers several states, and their diversity can be as far reaching as the diverse climate, soils, topography, and physiography of the entire United

States. Consequently, the approaches and solutions to water quality problems must be responsive to the unique characteristics of the surrounding watershed. EPA can set broad guidelines for each state's water quality program and can provide technical assistance in helping states meet the guidelines. There may be a leadership role for EPA on waterbodies that cross state boundaries, like the Chesapeake Bay. However, EPA cannot write and review all the designated uses that will apply to each of the nation's waterbodies, it cannot conduct all the monitoring and make all the listing decisions, and it cannot conduct the model analyses for all waterbodies. The scientific foundation for adaptive implementation must rely on state initiative and leadership. Today, EPA retains an extensive oversight role for the TMDL program. This raises the possibility that in an effort to ease the administrative burdens of reviewing and approving every TMDL, EPA will establish requirements for uniformity. This may result in standard setting, listing/delisting, and modeling approaches that are nationally consistent but are scientifically inappropriate for the planning and decision-making needs of the diversity of waterbodies. In the National Pollution Discharge Elimination System (NPDES) permitting program, EPA has helped states assume responsibility for point source permitting such that EPA does not review every permit that is issued. Using similar logic, EPA need not review every TMDL. The concern that the states cannot be relied upon to take action (Houck, 1999) needs to be tempered by the reality that continued extensive EPA oversight may not be feasible, it may place a premium on developing plans instead of taking actions, and it may inhibit the nation's progress toward improved water quality. The adaptive implementation approach may require increased state assumption of responsibility for individual TMDLs, with EPA oversight focused at the program level instead of on each individual water segment.

Conclusions and Recommendations

The call for adaptive implementation may not satisfy those who seek more definitive direction from the scientific community. Stakeholders and responsible agencies seek assurance that the actions they take will prove correct; they desire predictions of the costs and consequences of those actions in as precise terms as possible. However, waterbodies exist inside watersheds that are subject to constant change. For this reason and others, even the best predictive capabilities of science cannot assure that an action leading to attainment of designated uses will be initially identified. Adaptive implementation will allow the TMDL program to move forward in the face of these uncertainties.

1. EPA should act (via an administrative rule) to incorporate the elements of adaptive implementation into TMDL guidelines and regulations. To increase the scientific foundation of the TMDL program, the scientific method, which is embodied by the adaptive implementation approach, must be applied to water quality planning.

2. If Congress and EPA want to improve the scientific basis of the TMDL program, then the policy barriers that currently inhibit adoption of an adaptive implementation approach to the TMDL program should be addressed. This includes the issues of future growth, the equitable distribution of cost and responsibility among sources of pollution, and EPA oversight.

REFERENCES

Environmental Protection Agency (EPA). 1983. Technical Support Manual: Waterbody Surveys and Assessments for Conducting Use Attainability Analyses. Washington, DC: EPA Office of Water Regulations and Standards.

EPA. 1993. The Watershed Protection Approach, The Annual Report 1992. EPA 840-S-93-001. Washington, DC: EPA Office of Water.

EPA. 1994. Water Quality Standards Handbook: Second Edition. EPA 823-B-94-005a. Washington, DC: EPA Office of Water.

Houck, O. A. 1999. The Clean Water Act TMDL Program: Law, Policy, and Implementation. Washington, DC: Environmental Law Institute.

Novotny, V., J. Braden, D. White, A Capodaglio, R. Schonter, R. Larson, and K. Algozin. 1997. A Comprehensive UAA Technical Reference. 91-NPS-1. Alexandria, VA: Water Environment Research Foundation.

National Academy of Public Administration. 2000. Transforming Environmental Protection for the 21st Century. Washington, DC: National Academy of Public Administration. Page 86.

Reckhow, K. H. 1985. Decision Theory Applied to Lake Management. In: Proceedings of the North American Lake Management Society Conference, p. 196–200.

Robinson, A. R., and P. F. J. Lermusiaux. 2000. Overview of data assimilation. Harvard Reports in Physical/Interdisciplinary Ocean Science. Number 62. Cambridge, MA: Harvard University. 19p.

Appendix A
Guest Presentations at the First Meeting of the NRC Committee[15]
January 24–26, 2001

Introduction to the TMDL Program: Current Status and Future Plans
Don Brady, EPA Office of Water

Congressional Request for the Study—Senate
John Pemberton and Peter Washburn, Senate Committee on Environment and Public Works

Congressional Request for the Study—House
Susan Bodine, House Subcommittee on Water Resources and Environment

March 2000 GAO Report on Status of Water Quality Data
Patricia McClure, General Accounting Office

Environmental Perspective on the TMDL Program and this Study
Nina Bell, Northwest Environmental Advocates

State Perspectives on the TMDL Program and this Study
Robbi Savage, Association of State and Interstate Water Pollution Control Administrators
Shawn McGrath, Western Governors' Association

EPA's Pressing Science Issues for the TMDL Program
Lee Mulkey and Tom Barnwell, EPA Office of Research and Development

TMDL Case Studies
Bruce Zander, EPA Region VIII
Gail Mitchell, Bob Ambrose, and Tim Wool, EPA Region IV

---

[15] The NRC committee does not necessarily agree with all the comments or testimony given but all were taken into account.

Water Environment Research Foundation Support of TMDL Research
Dean Carpenter, Water Environment Research Foundation
Paul Freedman, Limno-Tech, Inc.
Kent Thornton, FTN & Associates

Stakeholder Presentations
Fred Andes, Federal Water Quality Coalition
Doug Barton, National Council of the Paper Industry for Air and
    Stream Improvement
Richard Bozek, Edison Electric Institute
Faith Burns, National Cattleman's Association
John Cowan, National Milk Producers Federation
Cynthia Goldberg, Gulf Restoration Network
Jay Jensen, National Association of State Foresters
Norman LeBlanc, Association of Metropolitan Sewerage Agencies
Mike Murray, National Wildlife Federation
Rick Parrish, Southern Environmental Law Center
Rob Reash, American Electric Power and the Utility Water Act Group
Dave Salmonsen, American Farm Bureau Federation

Appendix B
Biographies of the Committee Members and NRC Staff

Kenneth H. Reckhow *(chair)* is a professor at Duke University with faculty appointments in the School of the Environment and the Department of Civil and Environmental Engineering. In addition, he is director of The University of North Carolina Water Resources Research Institute and an adjunct professor in the Department of Civil Engineering at North Carolina State University. He currently serves as president of the National Institutes for Water Resources and is chair of the North Carolina Sedimentation Control Commission. He has published two books and over 80 papers, principally on water quality modeling, monitoring, and pollutant loading analysis. In addition, Dr. Reckhow has taught several short courses on water quality modeling and monitoring design, and he has written eight technical guidance manuals on water quality modeling. He is currently serving, or has previously served, on the editorial boards of *Water Resources Research, Water Resources Bulletin, Lake and Reservoir Management, Journal of Environmental Statistics, Urban Ecosystems, and Risk Analysis.* He received a B.S. in engineering physics from Cornell University in 1971 and a Ph.D. from Harvard University in environmental systems analysis in 1977. Dr. Reckhow is currently a member of the NRC's Committee to Improve the USGS National Water Quality Assessment Program.

Anthony S. Donigian, Jr., is president and principal engineer for AQUA TERRA Consultants. His expertise is in watershed modeling; nonpoint pollution and water quality modeling; chemical fate, transport, and exposure assessment; and model validation and testing. Mr. Donigian has 30 years of a broad range of experience in the development, testing, and application of modern analytical techniques for the assessment of environmental contamination and water resources planning problems.

He is an internationally recognized authority on modeling nonpoint pollution and chemical migration in the environment, primarily for water, soil, and groundwater systems. His recent research and applications studies have concentrated on regional and watershed-scale modeling of nutrients and impacts of management practices, movement of contaminants through the vadose zone, groundwater contamination by pesticides and hazardous wastes, model validation issues and procedures, and the evaluation of control alternatives such as best management practices, conservation tillage, and remedial actions at waste sites. Mr. Donigian received an A.B. in engineering sciences and a B.S. in engineering from Dartmouth College and an M.S. in civil engineering from Stanford University.

James R. Karr is a professor of aquatic sciences and zoology and an adjunct professor of environmental engineering, environmental health, and public affairs at the University of Washington, Seattle. He was on the faculties of Purdue University, University of Illinois, and Virginia Polytechnic Institute and State University; he was also deputy director and acting director at the Smithsonian Tropical Research Institute in Panama. He has taught and done research in tropical forest ecology, ornithology, stream ecology, watershed management, landscape ecology, conservation biology, ecological health, and science and environmental policy. He is a fellow in the American Association for the Advancement of Science and the American Ornithologists' Union. Dr. Karr has served on the editorial boards of *BioScience, Conservation Biology, Ecological Applications, Ecological Monographs, Ecology, Ecosystem Health, Freshwater Biology, Ecological Indicators,* and *Tropical Ecology.* He developed the index of biotic integrity (IBI) to directly evaluate the effects of human actions on the health of living systems. Dr. Karr holds a B.S. in fish and wildlife biology from Iowa State University and an M.S. and Ph.D. in zoology from the University of Illinois, Urbana-Champaign.

Jan Mandrup-Poulsen is an environmental administrator with the Watershed Assessment Section of the Florida Department of Environmental Protection. He is responsible for evaluating surface water quality, surface water/groundwater interactions, and mixing zones, and for determining the Total Maximum Daily Loads (TMDLs) allowable to support designated uses. He has coauthored materials on nonpoint source regulation in Florida and permitting guidance documents for point source discharges in Florida with consideration of the TMDL program. He is a frequent speaker on the topics related to the Florida Department of Environmental Protection watershed management approach, TMDLs, and the Impaired Waters Rule. Mr. Mandrup-Poulsen received his B.S. in atmospheric and oceanic science from the University of Michigan and his M.S. in biological oceanography and M.B.A. from Florida State University.

H. Stephen McDonald is a principal with Carollo Engineers. He has 22 years of experience in the areas of wastewater planning, watershed management, wastewater disinfection, biosolids treatment/reuse/disposal, and chemical and biological wastewater treatment/reuse. He is currently project manager for the development of TMDLs for several watersheds, including the Truckee River from Lake Tahoe to Pyramid Lake and the Calleguas Watershed in California. For the Truckee River, he is developing the Coordinated Monitoring Program and an adaptive management watershed/water quality modeling and stakeholder process to establish TMDLs for nutrients (nitrogen and phosphorus) and total dissolved solids (TDS). Mr. McDonald has developed master plans for water and wastewater treatment facilities in many western regions, including Sacramento County, the city of Fresno, CA; and the cities of Reno, Sparks, and Washoe County, NV. He holds a B.S. in biology from Portland State University and a B.S. in chemical engineering from Oregon State University. He has an MBA from California State University in Hayward and is a registered professional engineer in California.

Vladimir Novotny is a professor of environmental and water resources engineering at Marquette University and director of the Institute for Urban Environmental Risk Management. He is also president of the consulting firm Aqua Nova International, Ltd. His research has included risk-based urban watershed management integrating water quality and flood-control objectives, development of an adaptive methodology for online computerized modeling and real-time control of wastewater treatment facilities, and development of algorithms for control of urban sewer systems. He developed nationwide manuals on attainment of water quality goals (use attainability analysis) and abatement of winter diffuse pollution by road deicing operations. He is a past chair of an international group of specialists dealing with diffuse pollution and watershed management with the International Water Association. Dr. Novotny received a diploma engineer degree in sanitary engineering and a candidate of science degree in sanitary and water resources from the Technical

University of Brno, Czechoslovakia and a Ph.D. in environmental engineering from Vanderbilt University.

Richard A. Smith joined the Water Resources Division of the U. S. Geological Survey (USGS) in 1975 and began working with a small research team on statistical methods in water quality and their applications to the extensive and diverse water quality monitoring records maintained by the USGS. Throughout the 1980s, his research dealt with patterns of change in the nation's water quality and with statistical analysis of data collected from the more than 400 stream and river monitoring stations in the Survey's NASQAN program. In the early 1990s he began to investigate the possibility of using the rapidly advancing technology of GIS to enable the use of monitoring data in making statistically based predictions of water quality in unmonitored waters. For more than a decade he has also been very interested in the question of the adequacy of the nation's monitoring programs. He recently served on a panel of scientists charged with making recommendations for a comprehensive monitoring plan for the drinking-water supply watersheds serving New York City. Dr. Smith received his B.S. and M.S. in biology from the University of Richmond and his Ph.D. in environmental engineering from Johns Hopkins University.

Chris O. Yoder is manager of the Ecological Assessment Section of the State of Ohio Environmental Protection Agency. His current responsibilities include ecological evaluation of Ohio's surface water resources including streams, rivers, lakes, and wetlands; development of ambient biological, physical, and chemical assessment methods, indicators, and criteria for rivers, streams, inland lakes, wetlands, Lake Erie, and the Ohio River; reporting on the condition of Ohio surface water resources on a local, regional, and statewide scale; and development of environmental indicators for the surface water program. Previously he was a principal investigator of a cooperative agreement with the U.S. EPA Office of Water for developing approaches to implementing bioassessments and biological criteria within state and federal water quality management programs. Mr. Yoder received a B.S. in agriculture from Ohio State University and his M.A. in zoology from DePauw University.

Appendix B

NRC Staff

Leonard Shabman is a professor in the Department of Agricultural and Applied Economics at the Virginia Polytechnic Institute and State University and director of the Virginia Water Resources Research Center. He earned his Ph.D. in resource and environmental economics from Cornell University. His research interests include water supply, water quality, and flood hazard management; fishery management; and the role of economists in public policy formulation. Dr. Shabman was a member of the NRC's Committee on Watershed Management, Committee on USGS Water Resources Research, Committee on Flood Control Alternatives in the American River Basin, and the Committee on Restoration of Aquatic Ecosystems: Science, Technology, and Public Policy.

Laura J. Ehlers is a senior staff officer for the Water Science and Technology Board of the National Research Council. Since joining the NRC in 1997, she has served as study director for seven committees, including the Committee to Review the New York City Watershed Management Strategy, the Committee on Riparian Zone Functioning and Strategies for Management, and the Committee on Bioavailability of Contaminants in Soils and Sediment. She received her B.S. from the California Institute of Technology, majoring in biology and engineering and applied science. She earned both an M.S.E. and a Ph.D. in environmental engineering at the Johns Hopkins University. Her dissertation, entitled RP4 Plasmid Transfer Among Strains of *Pseudomonas* in a Biofilm, was awarded the 1998 Parsons Engineering/Association of Environmental Engineering Professors award for best doctoral thesis.